VOLUME ONE HUNDRED AND FORTY THREE

PROGRESS IN MOLECULAR BIOLOGY AND TRANSLATIONAL SCIENCE

Gonadotropins: From Bench Side to Bedside

VOLUME ONE HUNDRED AND FORTY THREE

PROGRESS IN MOLECULAR BIOLOGY AND TRANSLATIONAL SCIENCE

Gonadotropins: From Bench Side to Bedside

Edited by

T. Rajendra Kumar, PhD
Edgar L. and Patricia M. Makowski Endowed Professor, Department of Obstetrics & Gynecology, University of Colorado Denver-Anschutz Medical Campus, Aurora, CO, United States

ELSEVIER

AMSTERDAM • BOSTON • HEIDELBERG • LONDON
NEW YORK • OXFORD • PARIS • SAN DIEGO
SAN FRANCISCO • SINGAPORE • SYDNEY • TOKYO
Academic Press is an imprint of Elsevier

Academic Press is an imprint of Elsevier
125 London Wall, London EC2Y 5AS, United Kingdom
525 B Street, Suite 1800, San Diego, CA 92101-4495, United States
50 Hampshire Street, 5th Floor, Cambridge, MA 02139, United States
The Boulevard, Langford Lane, Kidlington, Oxford OX5 1GB, United Kingdom

First edition 2016

ISBN: 978-0-12-801058-7
ISSN: 1877-1173

For information on all Academic Press publications
visit our website at https://www.elsevier.com/

Publisher: Zoe Kruze
Acquisition Editor: Alex White
Editorial Project Manager: Helene Kabes
Production Project Manager: Magesh Kumar Mahalingam
Designer: Maria Ines Cruz

Typeset by Thomson Digital

CONTENTS

CONTRIBUTORS

S.L. Asa
Department of Laboratory Medicine and Pathobiology, University of Toronto, Toronto, ON, Canada; Department of Pathology, Laboratory Medicine Program, University Health Network, Toronto, ON, Canada

H.C. Blair
Departments of Pathology and of Cell Biology, University of Pittsburgh School of Medicine and the Pittsburgh VA Medical Center, Pittsburgh, PA, United States

G. Brigante
Unit of Endocrinology, Department of Biomedical, Metabolic and Neural Sciences, University of Modena and Reggio Emilia, Modena, Italy; Center for Genomic Research, University of Modena and Reggio Emilia, Modena, Italy; Azienda USL of Modena, Modena, Italy

L. Casarini
Unit of Endocrinology, Department of Biomedical, Metabolic and Neural Sciences, University of Modena and Reggio Emilia, Modena, Italy; Center for Genomic Research, University of Modena and Reggio Emilia, Modena, Italy

B.S. Ellsworth
Department of Physiology, School of Medicine, Southern Illinois University, Carbondale, IL, United States

S. Ezzat
Department of Medicine, University of Toronto, Endocrine Oncology, Princess Margaret Cancer Centre, University Health Network, Toronto, ON, Canada

J. Kapali
Department of Physiology, School of Medicine, Southern Illinois University, Carbondale, IL, United States

S. Lira-Albarrán
Department of Reproductive Biology, Instituto Nacional de Ciencias Médicas y Nutrición Salvador Zubirán, Mexico City, Mexico

P. Liu
The Mount Sinai Bone Program, Department of Medicine, and Department of Pediatrics, Icahn School of Medicine at Mount Sinai, New York, NY, United States

M. New
The Mount Sinai Bone Program, Department of Medicine, and Department of Pediatrics, Icahn School of Medicine at Mount Sinai, New York, NY, United States

T. Rajendra Kumar
Edgar L. and Patricia M. Makowski Endowed Professor, Department of Obstetrics & Gynecology, University of Colorado Denver-Anschutz Medical Campus, Aurora, CO, United States

D. Santi
Unit of Endocrinology, Department of Biomedical, Metabolic and Neural Sciences, University of Modena and Reggio Emilia, Modena, Italy; Center for Genomic Research, University of Modena and Reggio Emilia, Modena, Italy; Azienda USL of Modena, Modena, Italy

M. Simoni
Unit of Endocrinology, Department of Biomedical, Metabolic and Neural Sciences, University of Modena and Reggio Emilia, Modena, Italy; Center for Genomic Research, University of Modena and Reggio Emilia, Modena, Italy; Azienda USL of Modena, Modena, Italy

C.E. Stallings
Department of Physiology, School of Medicine, Southern Illinois University, Carbondale, IL, United States

L. Sun
The Mount Sinai Bone Program, Department of Medicine, and Department of Pediatrics, Icahn School of Medicine at Mount Sinai, New York, NY, United States

A. Ulloa-Aguirre
Research Support Network, Universidad Nacional Autónoma de México (UNAM)-National Institutes of Health, Mexico City, Mexico

T. Yuen
The Mount Sinai Bone Program, Department of Medicine, and Department of Pediatrics, Icahn School of Medicine at Mount Sinai, New York, NY, United States

M. Zaidi
The Mount Sinai Bone Program, Department of Medicine, and Department of Pediatrics, Icahn School of Medicine at Mount Sinai, New York, NY, United States

A. Zallone
Department of Histology, University of Bari, Bari, Italy

PREFACE

Knowing is not enough; we must apply. Willing is not enough; we must do.
Goethe

Basic and clinical research on pituitary gonadotropins started nearly 100 years ago. In the beginning, hypophysectomy, a surgical feat revealed the importance of pituitary hormones in many physiological systems including reproduction. Later, most of the focus was on whether two gonadotropins existed, shared a common alpha subunit that was linked to the hormone-specific beta subunit. Having realized that they did indeed exist and were heterodimers, the next goal was to develop specific and sensitive bioassays and immunoassays to measure them in circulation and pituitary extracts, and localize them within gonadotropes under a variety of physiological conditions. Along the way came the immunoneutralization approaches, which identified the specific need for LH and FSH in gonadal function. The localization of cell-surface receptors on gonads and their purification from gonadal cell membranes provided new insights into gonadotropin action. Subsequent structure-function studies laid the foundation for future three-dimensional modeling research. The above mentioned basic science discoveries slowly began to impact clinical research. Clinicians began testing the human urinary gonadotropins, albeit not entirely pure, on patients.

The advent of molecular biology and cloning of the subunit-encoding genes heralded a new era in gonadotropin gene regulation, and led to the production of pure, safe, and efficacious recombinant gonadotropic hormones for clinical use. Then came the major breakthrough. It was possible to achieve gene manipulation and understanding the genetics and physiology of gonadotropins at the whole organism level. This led to modeling human reproductive diseases (infertility and pituitary and gonadal tumors), in mice, and integrating the human patient data on polymorphisms and mutations in gonadotropins/their cognate receptors. These developments resulted in better diagnosis and designing treatment options for gonadotropin-dependent fertility disorders. Two major surprises came recently. The discovery of age-related FSH glycoforms and extragonadal FSH receptors. We must further explore the functional significance of these two controversial observations, because they have tremendous clinical significance, particularly, in ART protocols and menopause research. Unraveling the mysteries surrounding these two novel issues, may be a future goal in many research laboratories.

Volume 143 of the *Progress in Molecular Biology and Translational Sciences* (PMBTS) is devoted to *Gonadotropins: From Bench side to the Bedside.* Experts from all over the world have contributed chapters on Mouse Models for Gonadotrope Development (Chapter 1), Mouse Models for the Study of Synthesis, Secretion and Actions of Pituitary Gonadotropins (Chapter 2), Clinical Applications of Gonadotropins in the Female (Chapter 3), Clinical Applications of Gonadotropins in the Male (Chapter 4), Beyond Reproduction: Pituitary Hormone Actions on Bone (Chapter 5), and Gonadotrope Tumors (Chapter 6). I thank all the contributing authors for an excellent job of amalgamating the up to date knowledge on animal models and human conditions related to gonadotropins. These Chapters clearly illustrate how the bench side research work could benefit patients at the clinic. Undoubtedly, much remains to be done at both the frontiers—bench side and bedside on gonadotropin research. Certainly, there is a need and scope to further updating, including additional chapters, and bringing a new expanded volume in the future.

I thank Professor P. Michael Conn, PMBTS Series Editor, for inviting me to edit this state-of-the-art volume on gonadotropins. His constant support and genuine encouragement are truly inspiring. Finally, I owe my sincere thanks to Ms. Helene Kabes and her Production Team members at the Elsevier Press, for their patience, and rendering marvelous guidance and support throughout the journey. To all the Readers—enjoy the PMBTS, Volume 143, *Gonadotropins: From Bench side to Bedside.*

T. Rajendra Kumar, PhD
Editor

CHAPTER ONE

Mouse Models of Gonadotrope Development

C.E. Stallings, J. Kapali, B.S. Ellsworth[1]
Department of Physiology, School of Medicine, Southern Illinois University, Carbondale, IL, United States
[1] Corresponding author. E-mail address: bells@siu.edu

Contents

Abstract

The pituitary gonadotrope is central to reproductive function. Gonadotropes develop in a systematic process dependent on signaling factors secreted from surrounding

Progress in Molecular Biology and Translational Science, Volume 143
ISSN 1877-1173
http://dx.doi.org/10.1016/bs.pmbts.2016.08.001

tissues and those produced within the pituitary gland itself. These signaling pathways are important for stimulating specific transcription factors that ultimately regulate the expression of genes and define gonadotrope identity. Proper gonadotrope development and ultimately gonadotrope function are essential for normal sexual maturation and fertility. Understanding the mechanisms governing differentiation programs of gonadotropes is important to improve treatment and molecular diagnoses for patients with gonadotrope abnormalities. Much of what is known about gonadotrope development has been elucidated from mouse models in which important factors contributing to gonadotrope development and function have been deleted, ectopically expressed, or modified. This chapter will focus on many of these mouse models and their contribution to our current understanding of gonadotrope development.

1. INTRODUCTION

Central to reproductive function is the hypothalamic pituitary gonadal axis in which hypothalamic GnRH activates specific receptors on the surface of pituitary gonadotropes. Activation of GnRH signaling stimulates expression of the gonadotropin subunits and the GnRH receptor, *Gnrhr.*[1–16] The pituitary gonadotropins are dimeric glycoprotein hormones with a common α-subunit (*Cga*) and unique β-subunits (*Lhb* and *Fshb*) that give them their unique functions.[17] Gonadotropins are essential for gonadal function in both males and females.[18] Thus, the pituitary gonadotrope is vital for reproductive function.

The anterior lobe of the pituitary gland, together with the intermediate lobe, is derived from a structure referred to as Rathke's pouch. Rathke's pouch originates from oral ectoderm while the posterior lobe forms from neural ectoderm. During gestation most proliferating cells of Rathke's pouch border the luminal area. These cells then cease proliferating and migrate ventrally via an EMT-like transition to expand the anterior lobe. The anterior lobe has very few proliferating cells relative to the periluminal area.[19–22]

In vivo data suggest that pituitary cell specification occurs between embryonic day (e)10.5 and e12.5, while most pituitary cell types do not begin terminal differentiation until approximately e15.5.[23] Davis et al. used birth-dating studies to show that all anterior lobe cell types exit the cell cycle and begin the differentiation process between e11.5 and e13.5, suggesting that specialized cell types are not grouped together based on birth date.[24] At birth, the pituitary cell types are roughly organized into layers with gonadotropes being the most ventral. By adulthood spatial organization of the cell types appears more random, although recent studies demonstrate that the cell types form networks that are attached by adherens junctions.[25]

The layering of pituitary cell types at birth may be due to cell movement required to establish networks of specific cell types, rather than a relationship with the timing of cell cycle exit.[26]

Pituitary cell types express their signature hormones in a distinct temporal pattern. Hormone expression is dependent, in part, on regulation by specific transcription factors. The forkhead transcription factor, *Foxl2*, is coexpressed with *Cga*, the first hormone-encoding transcript to be detected initiating at approximately e10.5. CGA protein is present by e11.5.[27,28] The first gonadotrope-specific markers are *Nr5a1* and *Gnrhr* at approximately e13.5.[29,30] Birth-dating studies suggest that gonadotropes, which occupy a more rostral location during development than other anterior lobe cell types, exit the cell cycle and are specified in highest numbers at e11.5.[24] Although gonadotrope specification occurs early in the pituitary development, the gonadotropes terminally differentiate late in development with *Lhb* transcripts detectable by approximately e16.5 and *Fshb* shortly thereafter.[31] Gonadotropes are the least abundant of six hormone-producing cell types (gonadotropes, thyrotropes, somatotropes, lactotropes, corticotropes, and folliculostellate cells) in the anterior pituitary gland representing 5–10% of the anterior pituitary cells.[17]

There is increasing evidence that gonadotropes develop and persist as a heterogeneous population. Colabeling studies demonstrate the presence of two distinct gonadotrope subtypes at the beginning of gonadotrope differentiation: (1) LHB/GnRHR-positive cells and (2) FSHB/TSHB-positive, GnRHR-negative cells. The FSHB/TSHB-positive cells are thought to be the precursors of gonadotropes and thyrotropes. The FSHB-positive gonadotropes begin to express *Gnrhr* by e18.75.[55] By postnatal day (P)7, three distinct populations of gonadotropes exist: FSH-only gonadotropes, LH-only gonadotropes, and bihormonal gonadotropes with both FSH and LH (Fig. 1). While nearly all LHB-positive gonadotropes also contain NR5A1, only some FSHB-positive gonadotropes contain NR5A1.[33] Much effort has gone into understanding how undifferentiated progenitor cells become fully functional differentiated gonadotropes. In this chapter we will discuss many of the mouse models that have contributed to our understanding of gonadotrope development (Table 1).

2. SIGNALING PATHWAYS

Signaling factors that are intrinsic to Rathke's pouch, as well as factors secreted from the infundibulum, ventral diencephalon, and surrounding

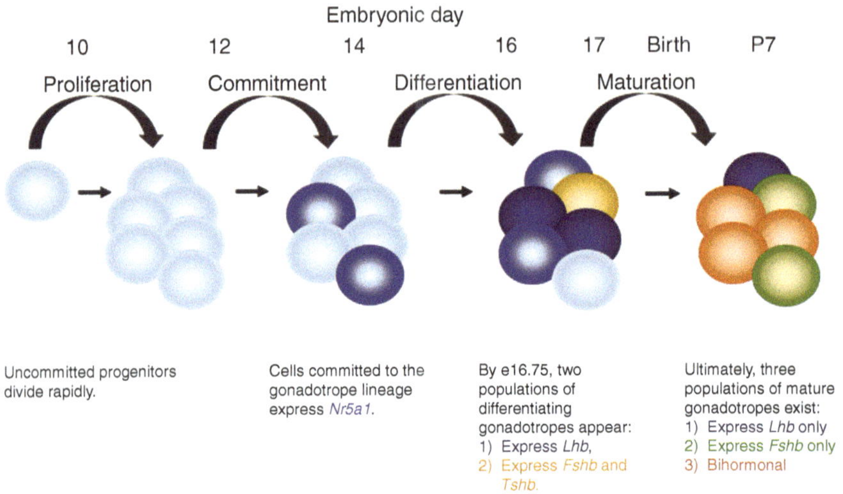

Figure 1 Model of gonadotrope development. During early pituitary development the cells of Rathke's pouch are rapidly dividing. Uncommitted progenitor cells, shown in *light blue*, contain PITX1, PITX2, and LHX3. These cells begin to commit to the gonadotrope lineage, *medium blue* cells, around e14.5 with the onset of *Nr5a1* and *Gnrhr* expression. These committed gonadotrope progenitors also contain GATA2.[23,68] Terminal differentiation begins around e16.5 with the expression of *Lhb*, *dark blue* cells. Studies by Wen et al. show that by e16.75 an occasional cell is positive for FSHB and TSHB, but not LHB *(yellow cells)*.[55] Ultimately, gonadotropes exist as a heterogeneous population with FSH-only gonadotropes *(green)*, LH-only gonadotropes *(dark blue)*, and bihormonal gonadotropes *(orange)*.[33] *P7*, Postnatal day 7.

mesenchyme regulate pituitary growth and morphogenesis and appear to be important for promoting differentiation of several pituitary cell types, including gonadotropes. Sonic hedgehog (SHH) and bone morphogenetic protein (BMP)2 are secreted from the ventral juxta–pituitary mesenchyme and diffuse into the surrounding tissue, including Rathke's pouch. Fibroblast growth factor (FGF)8 and BMP4 are secreted from the infundibulum creating a gradient of signaling factors in the developing pituitary gland.[96]

2.1 Fibroblast Growth Factors

Fgf8 is first detected in the infundibulum at e10.5.[96] Studies of pituitary explants show that FGF8 is important for the maintenance of *Lhx3* expression and repression of *Isl1* expression in the dorsal aspect of Rathke's pouch, suggesting that FGF8 signaling from the infundibulum is required to establish proper patterning of LIM homeobox gene expression during pituitary

Table 1 Mouse Models That Have Contributed to Our Knowledge of Gonadotrope Differentiation.

Genes	Description	Gonadotrope Phenotype	References
Signaling Pathways			
Fgf8	Hypomorphic allele, *Neo* insertion, global deletion, $Fgf8^{Neo/-}$	Variable, loss of anterior lobe to normal morphology with loss of LH	[34]
Fgf8	Hypomorphic allele, *Neo* insertion, $Fgf8^{neo/neo}$	Lack GnRH neurons at birth	[35,36]
Bmp4	Transgenic, *Cga-Bmp4*	Absence of *Cga*, *Tshb*, *Gh*, *Prl* at e17.0, dorsal expansion of expression domain for *Gata2*, *Isl1*, and *Msx1* at e17.0	[28]
Bmpr1a	Conditional deletion, $Bmpr1a^{flox/-}$; *Cga-cre*	Hypoplastic Rathke's pouch, loss of *Isl1* expression	[37,38]
Bmpr1a	Conditional deletion, $Bmpr1a^{flox/-}$; $Gnrhr^{+/GRIC}$	Fertile with normal gonadotropin levels in adulthood	[32,39]
Notch2	Transgenic, *Cga-Notch2*	Delayed gonadotrope differentiation	[40]
Notch2	Transgenic, *Pou1f1-Notch2*	Normal *Lhb* expression at e17.5	[41]
Notch2	Conditional deletion, $Notch2^{flox/flox}$;$Foxg1^{+/cre}$	Increased *Nr5a1* and *Lhb* expression at birth, but gonadotrope number is normal	[42–44]
Rbpjk	Conditional deletion, $Rbpjk^{flox/flox}$;*Pitx1-cre*	Normal expression of *Nr5a1* at e14.5, normal expression of *Lhb* at e17.5 suggests normal gonadotrope differentiation	[41]
Hes1	Transgenic, *Cga-Hes1*	Fewer gonadotropes and thyrotropes at birth	[45]
Shh	Global deletion, $Shh^{-/-}$	Pituitary gland is absent	[46]
Shh	Transgenic, *Cga-Shh*	Increased BMP2, expanded thyrotropes and gonadotropes	[46]
Hip	Transgenic, *Pitx1-Hip*	Inhibition of BMP signaling, severe reduction in size of Rathke's pouch	[46]

(*Continued*)

Table 1 Mouse Models That Have Contributed to Our Knowledge of Gonadotrope Differentiation.—cont'd.

Genes	Description	Gonadotrope Phenotype	References
Gli2	Global deletion of Zn-finger domain, *Gli2*$^{zfd/zfd}$	Variable loss of pituitary tissue, gonadotrope differentiation normal	[47]
Ctnnb1	Conditional deletion, *Ctnnb1*$^{flox(ex2-6)/flox(ex2-6)}$;*Gnrhr*$^{+/GRIC}$	Normal fertility	[32,48,49]
Ctnnb1	Conditional constitutive activation, *Ctnnb1*$^{flox(ex3)/+}$; *Gnrhr*$^{+/GRIC}$	Reduced LH and subfertility in males, females are fertile	[32,49,50]
Ctnnb1	Conditional constitutive activation, *Ctnnb1*$^{flox(ex3)/+}$; *Pitx1-cre* or *Pou1f1-cre*	Normal gonadotrope development	[51]
Gnrh	Spontaneous, *Gnrh*$^{hpg/hpg}$	Gonadotropes are present, but do not express *Lhb* or *Fshb* in the absence of exogenous GnRH	[4,52–54]
Gnrhr	Conditional ablation, *Rosa26-DTA;Gnrhr*$^{+/GRIC}$	Failure of FSH^+/TSH^+ pregonadotropes to differentiate into mature FSH-only gonadotropes	[32,55]
Transcription Factors			
Pitx1	Global deletion, *Pitx1*$^{-/-}$	Reduced gonadotrope population	[56]
Pitx1, *Pitx2*	Global deletion, *Pitx1*$^{+/-}$;*Pitx2*$^{+/-}$	Reduced gonadotrope population	[57]
Pitx2	Transgenic, *Cga-Pitx2*	Expansion of gonadotrope population	[58]
Pitx2	Hypomorphic allele, *Pitx2*$^{neo/neo}$	Absence of gonadotropes	[59]
Pitx2	Conditional deletion, *Pitx2*$^{flox-}$; *Lhb-cre*	Normal gonadotrope function	[60]
Lhx3	Global deletion, *Lhx3*$^{-/-}$	Reduced NR5A1 in the ventral pituitary and ectopic NR5A1 in the dorsal pituitary, dorsal pregonadotropes do not produce LHB	[61–63]
Lhx4	Global deletion, *Lhx4*$^{-/-}$	Reduced gonadotrope population	[64–66]

Isl1	Global deletion, $Isl1^{-/-}$	Hypoplastic, undifferentiated Rathke's pouch, embryonic lethality at e10.5	[67]
Gata2	Conditional deletion, $Gata2^{flox/flox}$;*Cga-cre*	Gonadotrope specification is normal, but the number of FSH-positive cells are reduced at birth	[68]
Gata2	Transgenic, *Pou1f1-Gata2*	Shift of precursor cells from thyrotrope to gonadotrope lineage	[23]
Gata2	Transgenic, *Gh-Gata2*	Normal gonadotrope differentiation	[23]
Gata2	Transgenic, *Cga-dnGata2*	Inhibition of terminal differentiation of gonadotropes and thyrotropes	[23]
Pou1f1	Transgenic, *Cga-Pou1f1*	Expansion of thyrotrope lineage into presumptive gonadotrope region	[23]
Pou1f1	Spontaneous, $Pou1f1^{dw/dw}$	Increased and dorsally expanded expression of gonadotrope markers	[23]
Pou1f1	Transgenic, $Cga\text{-}Pou1f1_{DBmut}$	Severe reduction in gonadotrope markers, thyrotropes normal	[23]
Prop1	Spontaneous, $Prop1^{df/df}$	Reduced gonadotrope population, persistent *Hesx1* expression, absence of *Notch2* expression	[65,69]
Prop1	Global deletion, $Prop1^{-/-}$	Reduced gonadotrope population, persistent *Hesx1* expression, absence of *Notch2* expression	[69]
Prop1	Transgenic, *Cga-Prop1*	Delayed gonadotrope differentiation resulting in delayed puberty and hypogonadism, normal NR0B1, NR5A1, *Gata2* and *Egr1*	[70]
Prop1	$Prop1^{df/df}$;$Lhx3^{-/-}$	Severe reduction of anterior lobe, expansion of NR5A1 domain	[65]
Hesx1	Global deletion, $Hesx1^{-/-}$	Variable, hypoplastic anterior lobe, animals with mildest phenotype were viable and fertile	[71]

(Continued)

Table 1 Mouse Models That Have Contributed to Our Knowledge of Gonadotrope Differentiation.—cont'd.

Genes	Description	Gonadotrope Phenotype	References
Hesx1	Compound heterozygote, *Six*$^{+/-}$; *Hesx1*$^{+/cre}$	Delayed gonadotrope differentiation, but normal specification	[72]
Otx1	Global deletion, *Otx*$^{-/-}$	Serum LH and FSH reduced 70–80% in adults, expression of *Gnrhr* and *Gnrh* is normal	[73]
Otx2	Conditional deletion, *Otx2*$^{flox/flox}$;*Pitx2*$^{+/cre}$	Normal number of CGA-positive cells, suggesting gonadotrope specification is normal	[44,74,75]
Pax6	Spontaneous, *Pax6*$^{Sey/Sey}$	Dorsally expanded domains of *Cga*, *Gata2*, *Prop1* and *Isl1* at e12.5 and *Nr5a1*, *Gata2* and *Foxl2* at e15.5, at birth expression of *Cga* is increased, *Lhb* is downregulated possibly due to lack of GnRH stimulation	[76,77]
Pax6	Global deletion, *Pax6*$^{-/-}$	Similar phenotype as *Pax6*$^{Sey/Sey}$	[76,78]
Pax6	Deletion of transactivation domain, *Pax6*$^{Neu/Neu}$	Similar phenotype as *Pax6*$^{Sey/Sey}$	[76,79,80]
Egr1	Global deletion, *Egr1*$^{-/-}$	Gonadotrope lineage is specified normally, but fail to produce LHB, FSHB is produced normally	[81]
Msx1	Global deletion, *Msx1*$^{-/-}$	Increased expression of *Gnrhr* and *Cga*, normal expression of *Lhb* and *Fshb* at e18.5	[82,83]
Tbx19	Global deletion, *Tpit*$^{-/-}$	Intermediate lobe cells differentiate into gonadotropes and thyrotropes	[84]
Tbx19	Transgenic, *Cga-Tbx19*	Severe reduction in the gonadotrope population	[84]
Nr5a1	Global deletion, *Nr5a1*$^{-/-}$	Gonadotropes are specified normally, express *Lhb* and *Fshb* in response to exogenous GnRH	[85]
Nr5a1	Conditional deletion, *Nr5a1*$^{flox/-}$; *Cga-cre*	Absence of *Lhb* and *Fshb* expression, produce LHB in response to exogenous GnRH	[86]

Nr0b1 (Dax1)	Global deletion, $Nr0b1^{flox/flox}$; *CMV-cre*	Normal number of LHB- and FSHB-positive cells	[87]
Foxl2	Transgenic, *Cga-Foxl2*	Stimulates ectopic expression of *Cga*	[27]
Foxl2	Global deletion, $Foxl2^{-/-}$	Reduction in *Fshb*, *Cga*, *Gnrhr*, and *Fst*, normal levels of *Lhb*, normal gonadotrope differentiation, increased pituitary cell density	[88,89]
Foxl2	Conditional deletion, $Foxl2^{flox/flox}$;$Gnrhr^{+/GRIC}$	Reduction in *Fshb* and *Fst*, normal levels of *Lhb*, *Cga*, *Gnrhr*	[32,90,91]
Foxd1	Global deletion, $Foxd1^{-/-}$	Reduced *Lhb* and normal *Fshb* expression at e18.5	[92]
Additional Genes			
Lsd1	Conditional deletion, $Lsd1^{flox/flox}$; *Pitx1-cre*	Reduced population of gonadotropes at e17.5	[51,93]
Dicer	Conditional deletion, $Dicer^{flox/flox}$;*Lhb-cre*	Reduced gonadotropin production in adult mice	[60,94,95]

GRIC, *Gnrhr*–internal ribosome entry site–*cre*.

development.[96] Human mutations in *FGF8* cause holoprosencephaly, diabetes insipidus, hypopituitarism, and, in some cases, hypogonadotropic hypogonadism.[35] Kallmann syndrome, characterized by an absence of functional GnRH neurons, can be caused by mutations in *FGF8* and *FGFR1*, the main receptor for mediating FGF8 signaling.[35,97,98] McCabe et al. employed a mouse model containing one hypomorphic allele of *Fgf8* ($Fgf8^{neo}$) and one null allele ($Fgf8^{-}$) to delineate the role of FGF8 in regulating pituitary development.[34,36] $Fgf8^{neo/-}$ mice have a variable phenotype with approximately one-third exhibiting a severe reduction of anterior pituitary and an absence of posterior pituitary at e17.5.[34] Approximately two-thirds of $Fgf8^{neo/-}$ mice have a milder phenotype with a morphologically normal pituitary, but a reduction or absence of *Lhb* expression. Other hormone-producing cell types were present.[34] $Fgf8^{neo/-}$ mice die immediately after birth, thus it is difficult to determine whether the absence of *Lhb* expression is due to loss of GnRH stimulation, or whether gonadotropes fail to differentiate.

Mice homozygous for the hypomorphic allele ($Fgf8^{neo/neo}$) lack GnRH neurons at birth.[35,36] These mice have a 55% reduction in functional FGF8 protein levels and die within 1 day of birth.[36] Mice that are heterozygous for the hypomorphic allele ($Fgf8^{+/neo}$) have approximately half the number of GnRH neurons as their wild type littermates at 120 days of age, although LH and GnRH peptide levels are normal. Females have delayed puberty and males have normal testicular development.[99] Establishing a mouse model with pituitary-specific deletion of *FGFR1* would greatly aid our understanding of the specific role of FGF8 in relation to gonadotrope development and differentiation.

2.2 Bone Morphogenetic Proteins

BMP signaling is critical for pituitary organogenesis.[28,37,96] *Bmp4* is expressed in the embryonic infundibulum and is required for the induction of Rathke's pouch.[67,96] *Bmp2* is expressed in the mesenchyme adjacent to Rathke's pouch at e10.5–e12.5 and throughout Rathke's pouch at e12.5.[37] Overexpression of *Bmp* in presumptive gonadotrope precursors results in dorsal expansion of the expression domain of the transcription factors, *Gata2*, *Isl1*, and *Msx1*, at e17.0 suggesting that expression of these factors in presumptive gonadotrope precursors is induced by BMP signaling.[23] Expression of *Cga*, *Tshb*, *Gh*, and *Prl* are almost entirely absent in *Cga-Bmp4* mice at e17.0.[28]

In vitro studies suggest that BMP2 and BMP4 stimulate *Fshb* expression by activating the type I receptor, BMPR1A.[100,101] Davis et al. conditionally deleted *Bmpr1a* in all pituitary cell types using *Cga-cre.*[37,38,102,103] At e10.5, Rathke's pouch is thin and underdeveloped in $Bmpr1a^{flox/-}$;*Cga-cre* embryos.[37] PITX1 and LHX3 are present in Rathke's pouch, suggesting that pituitary organogenesis has initiated. However, ISL1 is absent. Thus, induction of *Isl1* expression during pituitary organogenesis is dependent upon signaling through BMPR1A.[37] $Bmpr1a^{flox/-}$;*Cga-cre* exhibit early embryonic lethality at e12.5, likely due to *cre* expression and deletion of *Bmpr1a* in cardiac muscle causing heart defects.[37]

Surprisingly, deletion of *Bmpr1a* specifically in gonadotropes does not affect gonadotrope development or function.[32,39,102,103] $Bmpr1a^{flox/-}$; $Gnrhr^{GRIC/+}$ mice are fertile with normal serum gonadotropins and normal gonadotropin transcript levels.[39] The embryonic phenotype of these mice was not examined. The apparent inconsistency between these two mouse models is likely due to the fact that *Cga-cre* causes recombination in the pituitary primordium and all cell types very early in development with initial expression at e9.5, whereas $Gnrhr^{GRIC/+}$ causes stimulation only in gonadotropes and recombination occurs later at e12.75.[32] These studies suggest that BMP signaling is not required for gonadotrope function. However, additional studies are required to determine if BMP signaling is important for very early stages of gonadotrope development.

2.3 Notch

The Notch signaling pathway is key for the development of numerous tissues and has been implicated in several human developmental disorders.[104] The Notch receptor gene, *Notch2*, is expressed in Rathke's pouch from e12.5-e14.5 in the proliferating cells surrounding the lumen, but not in differentiated cells expressing *Cga.*[105] As pituitary development proceeds, *Notch2* expression decreases, consistent with a role in preventing cell differentiation. *Notch2* expression is absent in mice lacking the pituitary-specific transcription factor, *Prop1.*[105] Overexpression of the NOTCH target, *Hes1*, inhibits gonadotrope and thyrotrope differentiation.[45] Consistent with these findings overexpression of the constitutively active *Notch2* intracellular domain driven by the *Cga* promoter (*Cga-Notch2*) delays gonadotrope differentiation with almost no fully differentiated gonadotropes at birth. Differentiated gonadotropes do eventually develop, although these cells do not express the *Notch2* transgene.[105] In a similar study the *Notch2* intracellular domain was overexpressed under control of the *Pou1f1* promoter. In this scenario no effect was observed on *Lhb* expression at e17.5.[41] The difference in these studies may be due to the fact that *Cga* stimulates expression approximately 3 days earlier and in a more ventral domain than *Pou1f1*. Deletion of *Notch2* specifically in the pituitary gland (*Notch2*$^{fl/fl}$;*Foxg1*$^{+/cre}$) increases *Nr5a1* and *Lhb* expression at birth, although the number of gonadotropes is not statistically different from wild type littermate controls.[42–44] Similarly, conditional deletion of the NOTCH mediator, *Rbpj*, does not affect gonadotrope commitment, although corticotropes differentiate prematurely.[41] Together these data suggest that persistent NOTCH signaling inhibits gonadotrope differentiation, and although gonadotropin expression may be increased with loss of NOTCH signaling, gonadotrope development is unaffected.

2.4 Sonic Hedgehog

The morphogen, *Shh*, is expressed in the ventral diencephalon and the oral ectoderm, but its expression is extinguished from Rathke's pouch as it begins to form.[46] Treier et al. studied the role of SHH in pituitary gland differentiation. Global deletion of *Shh* (*Shh*$^{-/-}$) results in mice in which the pituitary gland is not detectable. This may be due to direct or indirect effects of SHH action.[46]

Using a transgenic approach, 9 kb of regulatory upstream sequence from the *Pitx1* gene was linked to the DNA sequence encoding *Hip* (*Pitx1-Hip*). HIP is an inhibitor of all mammalian hedgehog family members.[106] At e11.5 Rathke's pouch is hypoplastic and dysmorphic in *Pitx1-Hip* transgenics although *Lhx3* expression can be readily detected. By e13.0, the arrest of pituitary development is even more apparent. *Pitx1-Hip* mice display a pituitary phenotype similar to that of *Lhx3*$^{-/-}$ mice leading to speculation that *Lhx3* expression may be regulated by SHH in addition to FGF signaling.[46]

Complementary to loss-of-function studies, Treier et al. generated mice in which expression of the full-length coding region of *Shh* are regulated by the *Cga* promoter, targeting expression to Rathke's pouch. *Cga-Shh* transgenic mouse embryos exhibit significant pituitary hyperplasia at e17.5. Expression of *Lhx3* is slightly increased in the ventral aspect of the pituitary gland, again suggesting that *Lhx3* expression is stimulated by SHH. The increase in pituitary gland size is due to an increase in the number of *Cga*-expressing cells, consisting of both thyrotropes and gonadotropes, with premature onset of *Lhb* expression.[46] This demonstrates that SHH promotes gonadotrope and thyrotrope specification.

2.5 β-Catenin

Ctnnb1 (previously referred to as *Catnb*) is a gene involved in the canonical *Wnt* signaling proven to be active in pituitary development.[51] The protein product of *Ctnnb1*, CTNNB1 (also known as catenin beta-1 or β-catenin), is normally sequestered in the cytoplasm by a destruction complex, which promotes degradation of the molecule.[57] When various WNT ligands bind to their corresponding receptors, the pathway is stimulated and CTNNB1 translocates into the nucleus to induce transcription. Chromatin immunoprecipitation studies performed on wild type mouse pituitary gland reveal CTNNB1 associated with *Lef/Tcf*-binding regions on the *Axin2* promoter at e12.5, indicating a role in early pituitary gland development.[51] A number of mouse models with mutations in *Ctnnb1* expression have been developed and investigated for pituitary effects, primarily *Ctnnb1*$^{flox(ex2-6)/flox(ex2-6)}$ and *Ctnnb1*$^{flox(ex3)/+}$.

An allele for conditional deletion of *Ctnnb1* (*Ctnnb1*$^{flox(ex2-6)/flox(ex2-6)}$), was generated by flanking *Ctnnb1* exons 2–6 with *loxP* sites.[48] Researchers have used this model crossed with *Gnrhr* $^{+/GRIC}$ mice to induce disruption localized to gonadotropes and some neurons. *Ctnnb1*$^{flox(ex2-6)/flox(ex2-6)}$; *Gnrhr* $^{+/GRIC}$ mice have no observable phenotypic effects or changes in fertility despite verification of the recombination events.[32,49]

A complementary model with a dominant stable mutation was developed by flanking only exon 3 with *loxP* sites and is referred to as *Ctnnb1*$^{flox(ex3)/flox(ex3)}$ but was originally published as *Catnb*$^{lox(ex3)/lox(ex3)}$.[50] Upon CRE-mediated recombination the resultant offspring produce an in-frame mutated CTNNB1 truncated to eliminate the phosphorylation sites involved in ubiquitination and subsequent degradation, resulting in constitutive activation of CTNNB1 signaling.[50] Studies of heterozygous *Ctnnb1*$^{flox(ex3)/+}$ mice crossed with the aforementioned *Gnrhr* $^{+/GRIC}$ mice revealed significant effects on male offspring but not females.[32,50] Males analyzed at 8–12 weeks of age were subfertile with a slight decrease in circulating LH levels and reduced *Lhb* mRNA expression in the pituitary gland.[49] Additionally, pituitary *Fshb* mRNA was significantly decreased. By using primary cultures of pituitary cells from *Ctnnb1*$^{flox(ex3)/+}$;*Gnrhr*$^{+/GRIC}$ mice, researchers were able to stimulate *Fshb* mRNA production by treatment with activin. Further investigation revealed that primary cultures of pituitary cells from *Ctnnb1*$^{flox(ex3)/+}$;*Gnrhr*$^{+/GRIC}$ mice are more sensitive to inhibin stimulation than wild type controls. Overall, the authors concluded that *Ctnnb1* may contribute to male-specific *Fshb* mRNA production as part of the activin/inhibin/follistatin system.[50] Whether this is due to early developmental effects on gonadotrope specification or simply postnatal gonadotrope function is unknown.

Studies using cell culture and additional mouse models have implicated *Ctnnb1* as an activator of various transcriptional events related to gonadotrope development. *Pitx2*$^{+/-}$ mice injected with known *Wnt* pathway agonist lithium chloride exhibit increased *Pitx2* expression, again connecting CTNNB1 to early pituitary gland formation.[57] Cell culture experiments have shown CTNNB1 binds to NR5A1 as a necessary structure for further NR5A1/EGR1 binding and robust *Lhb* promoter activation.[107] These two processes are at opposite temporal ends of pituitary development; PITX2 regulates initial gonadotrope development (first detected at e8.5) and *Lhb* expression precedes terminal gonadotrope differentiation. Therefore, CTNNB1 may be a multifunctional factor in both gonadotrope specification and gonadotropin production.

Additional research has shown CTNNB1 to be a necessary factor in early *Pou1f1* activation using the *Ctnnb1*$^{flox(ex2-6)/flox(ex2-6)}$ mouse model mated to *Pitx1-cre* or *Pou1f1-cre*; however, surprisingly these mutants did not have significant alterations in gonadotrope development or function. The *Ctnnb1*$^{flox(ex2-6)/flox(ex2-6)}$;*Pitx1-cre* mice exhibit no *Pou1f1* expression in the anterior lobe past e14.5; however, this is an event not currently linked to gonadotropes.[51] Most recently a model using *Wnt1-cre* to delete *Ctnnb1* from the developing embryo demonstrates its crucial role in regulating signaling from the pituitary organizer to influence "pituitary gland growth, development, and vascularization".[108] While the data from these studies are not gonadotrope-specific, they do highlight upstream signaling events necessary for proper pituitary gland formation. Overall, mouse models with mutations in the *Ctnnb1* gene have provided data implicating CTNNB1 both in early pituitary developmental events upstream of *Pitx1* and *Pitx2* and directly preceding terminal differentiation. Data from murine models show constitutive *Ctnnb1* expression is associated with decreased fertility and gonadotropin levels specifically in male mice.

2.6 GnRH

GnRH secreted from parvocellular neurons of the hypothalamus into the pituitary portal vessel system regulates gonadotrope production of gonadotropins via binding to a seven transmembrane G-protein coupled receptor,

GnRHR.[1,5,17,109–112] *Gnrhr* is first detected in the developing gonadotrope at approximately e13.0, coincident with *Nr5a1*, making them the earliest gonadotrope-specific genes.[29,113] NR5A1 stimulates expression of *Gnrhr* via a complex enhancer.[114,115] Expression of *Gnrhr* is additionally stimulated by its own ligand, GnRH, via protein kinase C activation of an activator protein-1 (AP-1) element.[14–16,116] In vitro studies demonstrate several *Hbox* sites that contribute to basal and GnRH activation of the *Gnrhr* gene. Activation at these sites appears to involve, in part, binding of the homeobox transcription factor, OCT1.[117,118] The pituitary gland is able to secrete LH in response to GnRH as early as e16.[55] While GnRH is essential for normal expression of gonadotropins and *Gnrhr*, it does not appear to be required for differentiation of gonadotropes. This conclusion is based on studies of *hpg* mice, which do not produce GnRH due to a deletion in the *Gnrh* gene.[4,52] GnRH treatment stimulates gonadotropin production in *hpg* mice suggesting that gonadotrope differentiation occurs in the absence of endogenous GnRH.[52–54] Interestingly, ablation of *Gnrhr*-expressing cells appears to diminish the ability of FSHB/TSHB-positive pregonadotropes to differentiate into mature gonadotropes. The authors suggest that this is due to loss of LH production by the earlier arising LH-only gonadotropes. Therefore, GnRH signaling in late gestation may have subtle and indirect effects on gonadotrope subpopulations.[55]

3. TRANSCRIPTION FACTORS

Signaling factors play an influential role in regulating the expression of various transcription factors in the developing pituitary gland.[96,119] Many of these transcription factors are important for gonadotrope development. Some transcription factors are necessary for the earliest stages of pituitary development. Loss of these factors blocks not only development of the pituitary gland, but also terminal differentiation of one or more hormone-producing cell types. Mutations in transcription factors that are expressed specifically in gonadotropes can result in gonadotropin deficiency without any effects on the other pituitary cell types.

3.1 PITX1 and PITX2

Two of the earliest transcription factors expressed in the developing pituitary gland are the bicoid homeodomain factors, *Pitx1* and *Pitx2*. Initial expression of these factors is detected at e8.5 coincident with initiation of Rathke's pouch formation.[120,121] At this stage of development *Pitx1* and *Pitx2* are coexpressed in the uncommitted progenitor cells that make up the developing Rathke's pouch.[120,121] Later in adulthood *Pitx1* and *Pitx2* are expressed in all anterior pituitary cell types.[120,122]

As differentiated cell types begin to appear in the developing pituitary gland, *Pitx1* (also known as *Ptx1* and *P-Otx*) continues to be expressed in all hormone-producing pituitary cell types.[122,123] In vitro studies demonstrate that PITX1 can stimulate the expression of *Lhb* and *Lhx3*.[122,124,125] Global deletion of *Pitx1* in mice (*Pitx1*$^{-/-}$) results in an increased number of corticotropes and a decreased number of thyrotropes and gonadotropes at birth, suggesting that PITX1 is important for the balance of pituitary cell-type specification.[56] Interestingly, double heterozygous *Pitx1* and *Pitx2* mutant embryos demonstrate reduction of all the hormone-producing pituitary cell types at e18.5.[57]

PITX2 (PTX2, RIEG1) is present in differentiated thyrotropes, somatotropes, lactotropes, and gonadotropes, but not corticotropes.[120] Mutations in the human *PITX2* gene cause Rieger Syndrome. This is an autosomal dominant disorder characterized by ocular abnormalities, dental hypoplasia, craniofacial abnormalities, and, in some cases, decreased growth hormone levels.[121,126] Several mouse models have been established to analyze the dosage effects of PITX2. Global deletion of *Pitx2* (*Pitx2*$^{-/-}$) causes pituitary hypoplasia and lethality at e14.5 due to severe heart defects.[127–130] Reduction in *Pitx2* expression in mice homozygous for a hypomorphic allele of *Pitx2* (*Pitx2*$^{neo/neo}$) results in an absence of *Lhb*, *Fshb*, and *Gnrhr*. Other gonadotrope markers, including *Egr1* and *Nr5a1*, are also absent in these mice.[59,127] Overexpression of *Pitx2* in transgenic mice (*Cga-Pitx2*) results in the expansion of the gonadotrope population, as evidenced by an increased number of LH- and FSH-positive cells, which are present in a dorsally expanded domain at e18.5.[58] The expression domain for NR5A1, which is expressed in gonadotropes but no other pituitary cell types, was similarly expanded in *Cga-Pitx2* mice. No significant difference was observed in the number of somatotropes, thyrotropes, or corticotropes.[58] Together these studies show that gonadotrope differentiation is exquisitely sensitive to PITX2 dosage.

Interestingly, mice with gonadotrope-specific deletion of *Pitx2* (*Pitx2*$^{flox/-}$*;Lhb-cre*) have normal expression of LH and are fertile.[60,127] These data suggest that PITX2 is not required for gonadotrope maintenance or for regulated production of gonadotropes.[60] Many studies have shown that PITX2 transactivates the genes encoding the gonadotropin subunits: *Cga*, *Lhb*, and *Fshb*.[131,132] One possible explanation is that in the absence of PITX2, PITX1 may compensate to maintain gonadotrope function. Together these studies imply that PITX2 is more important in early gonadotrope development, while PITX1 is the most imperative for gonadotrope maintenance after birth.[60] Gonadotrope-specific deletion of both *Pitx1* and *Pitx2* would confirm the role of these factors in gonadotrope maintenance.

3.2 LIM Homeodomain Factors

The LIM homeodomain is named after the three members of this family of transcription factors: LIN1, ISL1, and MEC3.[133] In humans, mutations in the gene encoding the LIM protein, LHX3, are associated with combined pituitary hormone deficiency, which features GH, TSH, FSH, LH, PRL, and sometimes ACTH insufficiency. These patients have compound syndromes displaying dwarfism, hypogonadism, and hypothyroidism and often have rigid cervical spines, deafness, developmental delay, and intellectual disabilities.[134]

Sheng et al. studied mice with a global loss of *Lhx3* (*Lhx3*$^{-/-}$). Mice heterozygous for this *Lhx3* mutation are viable and fertile but homozygous embryos die shortly after birth. In the mutant embryos, induction of Rathke's pouch formation initiatess but its growth is arrested. Analysis of pituitary-specific lineage markers reveals that LHX3 is required for progenitor cells to commit to thyrotrope, gonadotrope, somatotrope, and lactotrope lineages. No LH-positive cells are present in e18.5 mutant pituitary.[61,62] Consistent with these findings, *Lhx3* null mice exhibit a reduction in the number NR5A1-positive cells in the ventral portion of Rathke's pouch at e14.5 and additional ectopic NR5A1-positive cells in the extreme dorsal aspect of the pituitary by e18.5. However, those cells do not stain for LH β-subunit, indicating that they are unable to terminally differentiate despite their expression of *Nr5a1*.[63]

Lhx4 is expressed throughout the invaginating pouch at e9.5, becomes restricted to the future anterior lobe of the pituitary gland at e12.5, and diminishes by e15.5.[64] Mice with homozygous *Lhx4* gene disruption have a severely reduced population of CGA-positive cells at e14.5 and at birth and few to no LH β-positive cells. The same is true for *Gnrhr* at both e15.5 and e18.5.[64] Thus, LHX3 is required for the cells of pituitary primordium to commit to gonadotrope lineage, as well as other pituitary cell types and LHX4 may support, but is not required for, specification of gonadotrope cells.

The LIM homeodomain factor, *Isl1*, is coexpressed with *Lhx3* early in pituitary development, but their expression domains become distinct by e12.5.[96] Specifically, *Isl1* is expressed throughout the oral ectoderm at e8.5, maintained in Rathke's pouch at e9.5, and ventrally restricted between e10.5 and e11.5.[96] By e12.5 it is expressed only in the ventral, differentiating cells that express *Cga* and *Foxl2*.[63,65] Global deletion of *Isl1* in mice causes arrested development soon after e9.5.[135] ISL1 and LHX3 are involved in early stages of pituitary ontogenesis. Together, these two factors illustrate the close relationship between the molecular mechanisms involved in cell differentiation and those involved in the expression of marker genes defining mature cell types.[136]

3.3 GATA2 and POU1F1

The GATA-binding family of transcription factors contains zinc fingers within their DNA-binding domains. *Gata2* is the most abundantly expressed GATA family member in the pituitary gland.[68] *Gata2* is expressed in the ventral pituitary during the closure of Rathke's pouch at e10.5. It is expressed at highest levels ventrally throughout pituitary development.[23] In adults, GATA2 is present in thyrotropes and gonadotropes.[68] Global deletion of *Gata2* is early embryonic lethal by e11.5 due to severe anemia.[68] In mice with pituitary-specific deletion of *Gata2* (*Gata2*$^{flox/flox}$;*Cga-cre*) the number of FSH-positive cells appear fewer at birth, although gonadotropes appear to have differentiated based on the presence of LH β-subunit immunoreactivity and NR5A1 expression.[38,68] In *Gata2*$^{flox/flox}$;*Cga-cre* adult mice, the number of FSHB-positive cells is normal, although serum FSH levels are reduced. Thus, gonadotropes are specified normally in the absence of GATA2, although GATA2 is important for normal gonadotrope function.[68]

Overexpression of *Gata2* (*Pou1f1-Gata2*) causes a reduction in the expression domain of dorsal pituitary cell types and an expansion of ventral cell types at e16.5. This phenotype persists at birth with an expansion of the gonadotrope lineage.[23] These data suggest that GATA2 is sufficient for shifting differentiation of precursor cells toward the gonadotrope lineage. While GATA2 is sufficient for specification of the gonadotrope lineage, gonadotropes are specified normally in *Gata2*$^{flox/flox}$;*Cga-cre* mice, suggesting that GATA2 is not required for this process.[68] This may be due to other transcription factors, such as GATA3 that may partially compensate for GATA2 during pituitary development.[68] This possibility is supported by studies of transgenic mice overexpressing a dominant negative form of *Gata2* (*Cga-dnGata2*).[23] In these mice, expression of GATA2 targets is repressed, effectively eliminating compensation by other factors. *Cga-dnGata2* embryos exhibit a severe reduction in *Lhb*, *Tshb*, *Cga*, and *Nr5a1*, but an expanded expression domain for *Pou1f1*.[23] These data suggest that GATA2 or other members of the GATA-binding protein family are necessary for gonadotrope differentiation.

The balance between expression of *Gata2* and *Pou1f1* appears to be important for appropriate gonadotrope specification.[23] POU1F1 (also known as PIT1) is required for differentiation of the POU1F1 lineage, which is comprised of thyrotropes, somatotropes, and lactotropes.[137] Misexpression of *Pou1f1* (*Cga-Pou1f1*, Table 1) in the ventral aspect of the pituitary results in expansion of thyrotropes and loss of gonadotropes at e18.5.[23] These data suggest that ectopic POU1F1 stimulates progenitor cells to switch their fate decision and differentiate into thyrotropes instead of gonadotropes (Fig. 1). Consistent with this finding, loss of POU1F1 function, as seen in $Pou1f1^{dw/dw}$ mice results in increased expression and dorsally expanded domains for *Lhb* and *Nr5a1* at birth.[23,137] In vitro studies demonstrate that POU1F1 can bind to GATA2 via protein–protein interactions and prevent GATA2 transactivation of *Nr5a1*.[23] Interestingly, POU1F1 does not inhibit binding of GATA2 to the *Tshb* promoter nor does it prevent transactivation of the *Tshb* promoter. In fact, they act synergistically.[23,138] This may explain how the balance between GATA2 and POU1F1 regulates gonadotrope and thyrotrope specification. This mechanism is supported by studies of mice that overexpress a form of POU1F1 incapable of binding DNA, but retaining the ability to interact with GATA2 ($Cga\text{-}Pou1f1_{DBmut}$).[23] These mice have normal *Tshb* and severely reduced *Lhb* and *Nr5a1* at e18.5.[23] Mutations of residues in POU1F1 that are necessary for interaction with GATA2 eliminate the ability of POU1F1 to reduce gonadotrope specification.[23] Together these studies confirm the suggestion that POU1F1 inhibits gonadotrope specification in a DNA-binding independent manner.

3.4 PROP1

Prop1 encodes a paired-like homeodomain transcription factor present in murine pituitary from about e10.5–e16.5. This gene has been the subject of intensive study, as humans with combined pituitary hormone deficiency frequently have mutations in the *Prop1* locus.[139] The name indicates its role as the Prophet of PIT1 and it was originally linked to the emergence of the

Pit-1 lineage of anterior lobe cells.[40] Recently it has been proven that *Prop1*-positive cells give rise to every other hormone-producing cell type in the anterior lobe of the pituitary gland.[140] This was effectively demonstrated using a *Prop1-cre* and lineage tracing experiments with the *Rosa26*^{tdTomatoEGFP} mouse reporter in which cells contain red fluorescent dtTOMATO protein in the absence of CRE activity or green fluorescent eGFP protein in the presence of CRE activity.[140,141]

The most extensively studied mouse model for *Prop1* function in embryonic development is the Ames dwarf mouse *Prop1*$^{df/df}$, first discovered as a spontaneous mutation causing dwarfism in a colony located in Ames, Iowa, USA. These mice have missense mutation Ser83Pro in the homeodomain of *Prop1*.[65,69] Although the mice exhibit no morphological changes until e12.5–e14.5, they eventually have anterior lobe defects, including ventral hypoplasia and dorsal dysmorphology.[20] *Prop1*$^{-/-}$ mice have essentially the same phenotype, although they occasionally die shortly after birth from respiratory distress.[69] The gonadotrope phenotype of these mice appears normal at 8 weeks of age, but reduced FSH β- and no LH β-subunits are detected at e18.5.[69] The Ames dwarfs exhibit reduced gonadotropin production, persistent *Hesx1* expression, a less defined region of *Nr5a1* and *Tle3*, reduced *Lef1* at e13.5–e14.5 in the anterior lobe, and no detectable *Notch2*.[40,51] The latter causes severe pituitary gland dysmorphology and is discussed further in the *Notch2* section of this chapter.[40] These data have led to the association of numerous signaling and transcription factors with *Prop1* and ultimately pituitary development.

Another model of study providing valuable data regarding the role of *Prop1* in gonadotrope development is the *Cga-Prop1* transgenic mouse, the most recent iteration being the TgN(*Cga-Prop1*)^{D65ac} strain.[70] Under control of the *Cga* promoter a 2.2-kb fragment of *Prop1* is expressed starting at e9.5 in the anterior lobe of the pituitary gland.[139] Multiple mice were generated and the model with the highest *Prop1* expression, as detected by in situ hybridization, was chosen for study. It was also determined that mutant mice on a C57BL/6 background were the most viable.[70,139] These PROP1 gain-of-function mice undergo delayed puberty and present with hypogonadism, most likely the result of a delay in gonadotrope differentiation, as *Lhb* and *Fshb* are not detected before birth.[70] Extensive study of this model has elucidated a novel mechanism mediating *Prop1* regulation of gonadotrope development as mutant animals had comparable NR0B1, NR5A1, POU1F1, *Gata2*, and *Egr1* at embryonic time points.[70] Microarray analysis identified *Prss28* and *Klrg1* as putative PROP1 targets responsible for the delayed puberty in gain-of-function animals. Unfortunately, no further study involving these genes has been published since 2006. Crossing the PROP1 gain-of-function mice with the Ames dwarf rescues the dwarf phenotype discussed earlier. Although this corrected the GH deficiency, the offspring maintained other defects including hypothyroidism.[139] Therefore, *Prop1* must be downregulated after e14.5 for normal cell differentiation to occur in the pituitary gland, and constitutive expression of the same results in delayed specification of gonadotropes.

Information on gonadotrope development has also been gathered from breeding the Ames dwarf with *Lhx4*$^{-/-}$ animals to obtain the *Prop1*$^{df/df}$; *Lhx4*$^{-/-}$ double knockouts. These mutants have almost no anterior lobe and at e12.5 no LHX3-positive cells.[65] Single mutants of *Prop1* or *Lhx4* maintain normal and reduced LHX3 levels, respectively, thereby indicating that *Prop1* plays some role in *Lhx3* expression in the pituitary gland. *Cga* is also not detected during embryogenesis in the double knockout model. There are no detectable hormone-expressing cell types at P0 with the exception of a reduced corticotrope population.[65] Interestingly, the NR5A1 domain is larger in the double-mutant pituitaries than in the *Lhx4* mutant where almost no NR5A1-positive cells exist. In *Prop1*$^{df/df}$ mice there is an expanded population of NR5A1-positive cells.[65] Therefore, PROP1 and LHX4 have antagonistic roles in regulating *Nr5a1*, with PROP1 repressing and LHX4 acting to upregulate expression.

3.5 HESX1

HESX1 is a homeobox protein expressed in the ventral diencephalon and Rathke's pouch of mice at e9.0. Its expression is extinguished by e15.5.[142] Mutations in the human *HESX1* gene cause septooptic dysplasia characterized by optic nerve hypoplasia, pituitary gland hypoplasia, and midline abnormalities of the brain.[143–146] Pituitary hypoplasia may manifest as isolated growth hormone deficiency or combined pituitary hormone deficiency consisting of reduction in GH, TSH, LH, and FSH.[147,148] Pituitary deficiency may be secondary to hypothalamic defects.[148] HESX1 acts as a transcriptional repressor and abnormalities with *HESX1* mutations may be due to the inability of HESX1 to interact with its corepressors Groucho/Transducin-like enhancer of split-1.[149,150]

Global deletion of *Hesx1* results in a variable phenotype.[71] The anterior pituitary is hypoplastic and less associated with the posterior lobe in all *Hesx1*$^{-/-}$ mice. In mice with the more severe class I phenotype, embryos lack a Rathke's pouch structure at e12.5.[71] The milder class II phenotype includes aberrant morphogenesis of Rathke's pouch in which the pouch contains bifurcations on each end at e12.5.[71] Approximately 1% of heterozygous *Hesx1* mutants exhibit the class II phenotype. While a number of class II mice die perinatally, some survive weaning. These mice were fertile and gave rise to viable offspring, suggesting that they have functional gonadotropes. There may be gene interactions that cause the variable phenotype. Overall, this study suggests that HESX1 is not essential for gonadotrope development or function in certain genetic contexts.[71]

SIX3 is a homeobox transcription factor that is required for normal forebrain formation.[151] The expression domains for *Six3* and *Hesx1* partially overlap within Rathke's pouch, suggesting they may both contribute to pituitary development. $Hesx1^{cre/+}$ mice were engineered by replacing the entire *Hesx1*-coding sequence with a *cre* cassette, thus creating a null allele for *Hesx1*.[72] These mice were mated to *Six3* heterozygous null mice to generate offspring that are doubly heterozygous for *Six3* and *Hesx1* ($Six3^{+/-};Hesx1^{cre/+}$).[72,151] $Six3^{+/-};Hesx1^{cre/+}$ mice exhibit two distinct phenotypes at e17.5. Two-thirds of $Six3^{+/-};Hesx1^{cre/+}$ embryos have hypoplastic pituitary glands with a reduced number of *Lhb*-positive cells. In one-third of the mutant embryos, the pituitary gland is absent from the sella and is instead ectopically located in the nasopharyngeal cavity. *Lhb* expression is barely detectable at e17.5 suggesting a delay in gonadotrope differentiation. Gonadotope specification is normal as evidenced by the expected expression patterns of *Nr5a1*.[72] These studies suggest that HESX1 and SIX3 contribute to normal *Lhb* expression and possibly gonadotrope differentiation.

3.6 OTX1 and OTX2

Otx1 encodes a homeobox transcription factor detected in mice from birth until approximately 4 months of age.[73] The *Otx1* gene is related to the *Ptx* gene family and in cell culture has been shown to bind the *Pitx1* recognition sequence to activate *Gh*, *Cga*, *Lhb*, and *Fshb* promoters, as well as interact with SMAD proteins.[73,152] Mice with *Otx1* exons 1 and 2 replaced by *lacZ* via homologous recombination were classified as $Otx1^{-/-}$.[153] Heterozygous mutants show no discernable phenotype therefore homozygous nulls were analyzed. Researchers used a hybrid B6/D2 background over the highly lethal 129/Sv, although the former still demonstrates 30% mortality in the first 30 days postbirth and a higher incidence of epileptic seizures.[73,153] $Otx1^{-/-}$ mice present with transient dwarfism from approximately P7 to 4 months of age, with the most severe size difference at P30. This is accompanied by a 70–80% reduction in circulating LH and FSH during the affected age. Testes and ovaries in mutant mice are significantly reduced in size during the dwarf period and males exhibit a decrease in mating frequency as compared to controls.[73] Pituitary gonadotropin expression as detected by immunohistochemistry shows a "strongly" reduced gonadotrope population at P25, which is restored to wild-type levels by 4 months of age.[73] Analysis of hypothalamic GnRH reveal no differences in protein levels compared to wild type mice, therefore any aberrant gonadotrope development is likely not the direct result of altered releasing hormone stimulation.[73] Histological analysis of *Gnrhr* expression and an apoptosis assay of the pituitary gland indicate normal levels of *Gnrhr* and imply no increase in programmed cell death.[73] The researchers did not treat mice with GnRH to verify competent hypothalamic signaling. However, by proving comparable numbers of *Gnrhr*-positive cells in the pituitary gland and GnRH expression in the hypothalamus, as well as no increase in cell death in the pituitary, they suggest that the gonadotrope population is intact in the mutant animals without normal LH and FSH expression.[73] Therefore, this model purports *Otx1* transiently regulates postnatal hormone production in the pituitary as opposed to gonadotrope development, potentially in concert with SMAD proteins.

Patients with mutations in *OTX2* exhibit a spectrum of phenotypes that include, in some instances, combined pituitary hormone deficiency and hypogonadotropic hypogonadism.[34] Global deletion of *Otx2* in mice results in embryonic lethality due to absence of forebrain, midbrain, and rostral hindbrain.[154] Conditional deletion of *Otx2* using *Foxg1*$^{cre/+}$ mice resulted in an array of pituitary dysmorphology phenotypes, but did not affect cell specification based on CGA and POU1F1 immunohistochemistry.[44,74,75] Similarly, *Otx2*$^{flox/flox}$;*Pitx2*$^{+/cre}$ mice exhibited proper cell specification and pituitary morphology was typical.[74,155] These studies suggest that gonadotrope specification is not dependent on OTX2. Gonadotrope phenotypes in human patients may be due to loss of hypothalamic function or an inability to maintain gonadotrope function.

3.7 PAX6

Pax6 is a transcription factor that is first detected at e9.0 with the protein evident by e11.5. It is primarily localized to the dorsal portion of the developing pituitary gland.[76] Expression is observed in the diencephalon by e9.5, a structure referred to as the pituitary organizer.[78] Various mouse models have greatly aided the study of *Pax6* originating with the spontaneously mutated *Small eye* or *Sey* mice ($Pax6^{Sey/Sey}$) and a recapitulation of the same designated as $Pax6^{-/-}$.[76–78] $Pax6^{Sey/Sey}$ mice have a point mutation in chromosome 2 truncating the translated protein and resulting in functionally inactive PAX6.[79,156] A number of models similar to $Pax6^{Sey/Sey}$ have been developed including $Pax6^{-/-}$ and $Pax6^{Neu/Neu}$ (the latter maintains the homeobox domain but lack the transactivation domain) and because of comparable pituitary phenotypes these three models will be discussed as a whole.

Absence of functional PAX6 results in mice with no GnRH-immunoreactive neurons during development.[80] Thus, any gonadotrope development would be independent of GnRH stimulation. Pituitary glands of PAX6-deficient mice exhibit ventralization, a shift in the identity of cells where dorsal pituitary cells adopt ventral fates.[76] In wild type mice gonadotropes are located primarily in the ventral aspect of the pituitary gland, and an increase of this cell type is observed in $Pax6^{-/-}$ mutants as ventralization occurs. Additionally, no LHB-positive cells in the CGA-positive population were observed reaffirming previous studies connecting *Pax6* to hypothalamic GnRH stimulation.[76] Overall, this model emphasizes the effect of spatial development signals on controlling the specification of the hormone-producing cell type make-up of the pituitary gland, and the precise regional control *Pax6* has over the gonadotrope lineage. Whether the alteration in gonadotrope function is entirely due to the absence of GnRH stimulation from the hypothalamus or is also effected by ventralization is unclear. Studies of *Pax6* knockout mice with the mutation localized specifically to either the pituitary gland or the hypothalamus would be valuable to understand this phenomenon.

3.8 EGR1

EGR1, also known as KROX24, NGFIA, and ZIF268, is an immediate early gene containing a DNA-binding domain composed of three zinc-finger motifs.[157,158] EGR1 is expressed in the endothelium, thymus, muscle, cartilage and bone, and the central and peripheral nervous systems.[159,160] EGR1 is important for numerous physiological processes including synaptic plasticity, wound repair, inflammation, and differentiation.[161] In vitro studies suggest that EGR1 synergizes with PITX1 and NR5A1 to stimulate expression of *Lhb* in response to GnRH.[125,162,163] In the pituitary gland, *Egr1* is expressed in the posterior and anterior lobes, as well as in the mesenchyme surrounding the pituitary gland during development. In the adult, *Egr1* expression is absent from the posterior pituitary, but strongly expressed throughout the anterior lobe.[81] Global deletion of *Egr1* (*Egr1*$^{-/-}$) results in reduced growth rate of mice. Both male and female mutants are infertile with pituitary and ovarian defects. Specifically, pituitary somatotropes are reduced in number and serum GH levels are low at 12 weeks of age. Gonadotropes are present in expected quantities based on the number of FSH β subunit–positive cells. However, LH β-subunit is absent and serum LH is undetectable.[81] Ovariectomy of *Egr1*$^{-/-}$ mice stimulates *Fshb*, but not *Lhb* expression.[162] Interestingly, expression of the gene encoding the LH receptor, *Lhcgr*, is absent in the ovary.[81] Based on these studies it appears that ERG1 is required for normal expression of *Lhb*, but not for differentiation of the gonadotrope lineage.

3.9 MSX1

Msx1, originally known as *Hox7*, is a homeobox gene with similarity to muscle segment homeobox (*msh*) from *Drosophila*.[164] *Msx1* is expressed in facial processes of e10.5–e13.5 mouse embryos, as well as the ventral aspect of the developing pituitary gland.[165,166] *Msx1* is first expressed in the pituitary gland at e10.5 becoming localized to ventral aspect of Rathke's pouch by e12.5.[166,167] The ventral pattern of *Msx1* expression together with the fact that it is expressed in gonadotrope-derived cell lines suggests that MSX1 may play a role in gonadotrope development or function.[167] A mouse model for global deletion of *Msx1* was generated by insertion of a *lacZ* cassette into the homeobox domain of *Msx1*.[82,83] *Msx1*$^{-/-}$ mice exhibit perinatal lethality with defects in craniofacial development, cleft palate, and tooth agenesis.[168,169] These mice exhibit increased *Gnrhr* and *Cga* expression at e18.5.[167] *Lhb* and *Fshb* are not significantly increased, but show a trend upward.[167] This is consistent with in vitro studies demonstrating the ability of MSX1 to repress *Gnrhr* and *Cga* promoter activity.[167] MSX1 may be important for regulating timing of gonadotrope differentiation.[167] Inhibition of BMP signaling by overexpressing a dominant negative version of *Bmpr1a* (*Cga-ΔBMP*) stimulates robust upregulation of *Msx1* expression throughout the pituitary gland at e17.0. Surprisingly, gonadotropes are specified despite increased expression of *Msx1*.[28] Studies that examine the timing of gonadotrope differentiation in *Msx*$^{-/-}$ mice will be important to determine whether deletion of *Msx1* causes premature differentiation of gonadotropes.

3.10 TBX19

The T-box transcription factor, TBX19 (also known as TPIT), was identified as a cell-specific factor that stimulates expression of the pituitary *Pomc* gene. TBX19 is first detected in nuclei of pituitary corticotropes at e12.5. At e13.5, all corticotropes contain TBX19.[170] In mice with a global deletion of *Tbx19* in which most of the *T box*–coding sequences are deleted (*Tbx19*$^{-/-}$), the pituitary exhibits apparently normal histology at e14.5 but the number of POMC-positive cells is greatly reduced.[84] These cells may be blocked in their differentiation process or may differentiate into other cell types. In the absence of TBX19, a large number of NR5A1-positive cells are ectopically located in the intermediate lobes of *Tbx19*$^{-/-}$ mice at e16.5, suggesting that progenitor cells of the intermediate lobe have erroneously differentiated into gonadotropes.[84] Colocalization experiments show that intermediate lobe cells of *Tbx19*$^{-/-}$ mice express *Pomc* or *Cga*, never both, and that *Cga* and *Lhb* expression is colocalized in many cells.[84] A few POU1F1-independent thyrotropes are also present in the intermediate lobe. Thus in the absence of TBX19, the intermediate lobe contains gonadotropes, POU1F1-independent thyrotropes, as well as the expected cell type, melanotropes. These three cell types do not appear to have a mixed or abnormal cell identity.[84] Cell fate changes also occur in the anterior lobe. The caudal region of the anterior lobe of *Tbx19*$^{-/-}$ mice contain NR5A1at e16.5. At this developmental time point, this caudal part of the pituitary does not normally have NR5A1- or CGA-expressing cells, suggesting a cell fate switch from the corticotrope lineage to the gonadotrope lineage.[84]

Gain-of-function experiments using the *Cga* promoter to drive *Tbx19* expression in the gonadotrope lineage demonstrate that pituitaries of such transgenic mice have more POMC-positive cells and fewer CGA-positive cells in the anterior lobe while TSHB-positive cells are unaltered. LH β-subunit is undetectable and FSH β-subunit is significantly decreased. Also, very few NR5A1-positive cells are present.[84] These data suggest that TBX19 is a negative regulator of gonadotrope differentiation.

3.11 Orphan Nuclear Receptors

Nr5a1 (also known as *Sf1* or *Ftz-F1*) is expressed in pituitary gonadotropes, as well as hypothalamus and steroidogenic tissues.[171] The onset of pituitary expression of *Nr5a1* occurs at approximately e13.5.[30] *Nr5a1* regulates many of the genes that constitute the gonadotrope phenotype including *Gnrhr*, *Lhb*, and *Fshb*.[114,162,172]

Mice with homozygous disruption of the *Nr5a1* gene (*Nr5a1*$^{-/-}$) lack detectable LH, FSH, and GnRHR. However, GnRH treatment restores gonadotropin expression.[30] Similarly, mice with pituitary-specific deletion of *Nr5a1* (*Nr5a1*$^{flox/-}$*;Cga-cre*) have severely reduced expression of gonadotropins and GnRH treatment restores production of LH.[86] Together these studies suggest a role for NR5A1 in normal gonadotrope function, but not differentiation.

Another orphan nuclear receptor implicated in gonadotrope development and function is NR0B1. Human males with mutations in the gene *NR0B1* (also known as *DAX1* and *AHCH*) exhibit hypogonadotropic hypogonadism.[87] In an attempt to recapitulate this phenotype, a mouse model was created by crossing *Nr0b1*$^{flox/flox}$ mice with *CMV-cre* mice.[87] The pituitary glands isolated from these mice had no change in LH β- and FSH β-subunit staining as compared to controls.[87] Additional testing of these mutants on a mixed B6/129Sv genetic background revealed aberrant gonadal differentiation. However, pituitary-specific deletion of *Nr0b1* was not performed nor was *Nr5a1* measured.[173] Molecular studies implicate *Nr0b1* and *Nr5a1* as negative regulators of each other, but no in vivo experiments have yet revealed a specific role for *Nr0b1* in gonadotrope development or gonadotropin production.[107,174]

3.12 Forkhead Box Transcription Factors

Forkhead transcription factors are classified based on a conserved winged helix/forkhead DNA-binding domain consisting of three β-sheets, three α-helices, and two loops forming the "wings".[175] FOXL2, is present in the pituitary, ovary, and developing eye and is the forkhead factor that has been studied most extensively in reproduction.[27,88–91,176–186] *Foxl2* transcript is first detected in the developing pituitary at e10.5 and FOXL2 protein is detectable a day later at e11.5.[27,28] FOXL2 appears to serve as a marker of gonadotrope and thyrotrope commitment.[28] In adult pituitary FOXL2 is present in gonadotropes and thyrotropes and during pregnancy, FOXL2 is present in a small fraction of lactotropes.[27] FOXL2 interacts with SMADs to stimulate expression of gonadotrope genes.[116,177,178,183]

Overexpression of *Foxl2* in pituitary tissue (*Cga-Foxl2*) is sufficient to stimulate production of CGA. Consistent with this, FOXL2 protein is first detectable at e11.5 in the rostral tip of the developing pituitary, coincident with the spatial and temporal patterns of CGA protein.[27] Interestingly, in *Cga-Foxl2* embryos ectopic FOXL2 was also detected in discrete regions bordering the posterior lobe and was sufficient to drive ectopic expression of CGA in these regions as well.[27] In vitro studies suggest that FOXL2 also stimulates expression of *Gnrhr*.[116] Pituitary glands from mice with a global deletion of *Foxl2* (*Foxl2*$^{-/-}$) have very low levels of *Fshb* and *Cga* and reduced levels of *Gnrhr* and *Fst*.[89] Activin is unable to stimulate *Fshb* transcription, suggesting that FOXL2 mediates activin stimulation of *Fshb* expression.[88] Gonadotrope differentiation does occur as evidenced by the presence of LH.[88] Pituitary glands from mice with gonadotrope-specific deletion of *Foxl2* (*Foxl2*$^{flox/flox}$;*Gnrhr* $^{+/GRIC}$) are of normal size with reduced expression of *Fshb* and *Fst*, although expression of *Cga*, *Lhb*, and *Gnrhr* is typical.[32,90,91] This difference in these mouse models may be due to the distinct timing of *Foxl2* deletion in the two models or to the absence of *Foxl2* in nonpituitary tissues in *Foxl2*$^{-/-}$ mice.

The forkhead transcription factor, FOXD1 (also known as BF2), is required for normal kidney development.[187,188] *Foxd1* is expressed in the mesenchyme surrounding the pituitary gland, but not in the pituitary gland itself during development.[92] Like many forkhead factors, FOXD1 is a single-exon gene, thus nearly the entire coding sequence was replaced with a *lacZ* cassette to generate a null allele.[187] *Foxd1*$^{-/-}$ mice die shortly after birth due to renal failure.[187] *Foxd1*$^{-/-}$ mice have a dysmorphic pituitary gland that extends through the cartilage plate from e14.5 through e18.5.[92] Expression levels of *Gh*, *Tshb*, *Pomc*, *Fshb*, and *Cga* are normal at e18.5. However, in spite of the fact that *Foxd1* is not expressed in the pituitary during development, *Lhb* expression is significantly reduced in embryos lacking *Foxd1*.[92] These findings indicate that FOXD1 is required for normal gonadotrope function during development. Based on its expression in the mesenchyme surrounding the pituitary gland, FOXD1 likely acts indirectly via signaling factors secreted from the mesenchyme.

3.13 Additional Genes Known to Contribute to Gonadotrope Development

While much is known about regulation of the genes involved in gonadotrope development by classical transcription factors, little information is currently available regarding nonclassical methods controlling gonadotrope development and function. Epigenetic regulation involves stable, heritable alterations of the genome that do not affect DNA sequence. Epigenetic regulation is essential to normal development and differentiation.[189] Two basic mechanisms of epigenetic regulation are DNA methylation and histone modification. Lysine-specific demethylase 1 (LSD1) is a histone lysine demethylase.[190,191] Global deletion of *Lsd1* results in early embryonic lethality at e7.5.[93] Pituitary specific deletion of *Lsd1* (*Lsd1*$^{flox/flox}$*;Pitx1-cre*) reveals a role for LSD1 in terminal differentiation of pituitary cell types.[51,93] *Lsd1*$^{flox/flox}$*;Pitx1-cre* embryos exhibited a severe reduction in markers for terminal differentiation of somatotropes, thyrotropes, gonadotropes and corticotropes. *Lhb* expression was barely detectable and *Nr5a1* expression was diminished at e17.5. *Cga* expression was apparent, although very few CGA-positive cells could be detected by immunohistochemistry, suggesting possible posttranscriptional regulation.[93] Thus, epigenetic regulation is an important component of gonadotrope development.

The amount of protein produced depends not only on the rate of transcription, but also on transcript stability and the rate of translation into protein. In vitro studies suggest that GnRH regulates stability of *Lhb*, *Cga*, and *Gnrhr* transcripts.[192–198] Translation of transcripts can be inhibited through binding of short, ~22 nucleotide RNAs called microRNAs. These microRNAs are processed from primary transcripts via a multistep process involving a cytoplasmic double-stranded, RNA-specific RNase called *Dicer*.[199] To evaluate the role of microRNAs in gonadotropes, Wang et al. deleted *Dicer* specifically in gonadotropes (*Dicer*$^{flox/flox}$*;Lhb-cre*).[60,94,95] Mice lacking *Dicer* in gonadotropes had significantly reduced expression of *Fshb*, *Lhb*, and *Cga*. Also, stimulation of LH production by GnRH was abolished in the absence of *Dicer*.[94] These studies suggest that posttranscriptional regulation is an important mechanism controlling gonadotropin production. Use of an earlier acting CRE would allow researchers to determine whether epigenetic regulation is equally vital during gonadotrope differentiation.

4. CRE MICE FOR TARGETING GONADOTROPES

Cre-loxP strategy allows for disruption of genes without embryonic lethality caused by global knockouts and eliminates secondary effects due to deletion of genes in other tissues. Specific promoters to regulate the temporal and spatial expression of *cre* are key to this process. Several CRE mouse lines have been developed that target gonadotrope cells (Table 2).

The gene encoding the receptor for GnRH, *Gnrhr*, is one of the earliest gonadotrope-specific genes to be expressed, making it an excellent choice for gonadotrope-specific expression studies in embryos and adults.[29]

Table 2 CRE Mice That Target Gonadotropes.

CRE Line	Generation	Targeted Expression	References
GRIC MGI ID: 3795249	Targeted mutation, IRES and *cre* inserted in frame downstream of exon 3 of *Gnrhr*	Recombination in gonadotropes and gonadotrope precursors, as well as GnRH neurons, recombination detected by e12.75	[32,55,201]
Cga-cre MGI ID: 2179995	Transgenic, 4.6 kb of mouse *Cga* fused to *cre*	Recombination in pituitary primordium and all five pituitary cell types, some ectopic CRE activity recombination by e9.5	[37,38]
Cga-cre MGI ID: 3842442	Transgenic, 4.5 kb of mouse *Cga* fused to *cre*	Recombination in pituitary gland, no recombination detected in hypothalamus	[202]
Cga-icre MGI ID: 5547616	Transgenic, 4.6 kb of proximal promoter from mouse *Cga* fused to *cre* with reduced susceptibility to epigenetic silencing	CRE activity in adult gonadotropes and thyrotropes with high penetrance, no ectopic CRE activity, recombination in differentiating pituitary cells at e12.5	[203]
Bα/CreTeR MGI ID: NA	Transgenic, –313 to +48 of bovine *Cga* driving tetracycline-inducible expression of *cre*	Recombination occurs in adult gonadotropes upon doxycycline treatment, recombination in other cell types is undetermined	[205]
Lhb-cre MGI ID: 3822455	Transgenic, 776 bp of bovine *Lhb* promoter fused to *cre* with engineered KOZAK	Gonadotrope-specific, recombination by e16.5	[60]
Fshb-icre MGI ID: NA	Transgenic, 4.7 kb of proximal promoter from ovine *Fshb* fused to *cre* with reduced susceptibility to epigenetic silencing	CRE activity beginning embryonically, detected in 99–100% of adult gonadotropes, no recombination in hypothalamus, testis, or ovary	[209]
Nr5a1-cre MGI ID: 3689423	Transgenic, BAC containing –23 kb to +88 kb of mouse *Nr5a1* with *cre* inserted at ATG initiating methionine	Recombination in pituitary by e13.5, recombination also in somatic cells of gonads, adrenal cortex, spleen, ventromedial hypothalamic nucleus, choroid plexus, hippocampus, and cerebral cortex	[211]

Foxl2$^{tm1(GFP/cre/ERT2)Pzg}$ MGI ID: 4940388	Targeted mutation, GFP and a tamoxifen-inducible form of *cre* replaced the single-coding exon for *Foxl2*, creates a null allele of *Foxl2*	Recombination in pituitary not examined, recombination detected in ovary	[212]
Foxl2$^{tm1.1(cre/ERT2)Kliu}$ MGI ID: 5548141	Targeted mutation, IRES and tamoxifen-inducible *cre* were inserted downstream of the *Foxl2* gene	Recombination in pituitary not examined, recombination detected in ovary	[213]
Foxl2$^{tm4(cre)Tre}$ MGI ID: 4430207	Targeted mutation, *cre* replaced the translational start site of *Foxl2*, creates a null allele of *Foxl2*	Recombination in pituitary not examined, recombination detected in ovary	[91]
ELHX3-cre MGI ID: 5576209	Transgenic, 7.9-kb enhancer from human *LHX3* and *Hsp68* minimal promoter fused to *cre*	Recombination in caudomedial pituitary gland at e14.5, recombination in CGA-, TSHB-, FSHB-, and LHB-positive cells at birth	[33]
Isl1-cre MGI ID: 3623159	Targeted mutation, *cre* inserted into *Isl1* locus replacing endogenous ATG	Recombination in pituitary not examined, recombination detected in cardiac and hindlimb progenitors	[219]
Isl1-IRES-*cre* MGI ID: 2447758	Targeted mutation, IRES and *cre* inserted downstream of the exon containing the second LIM domain of *Isl1*	Recombination in pituitary not examined, recombination detected in motor neurons and dorsal root ganglia at e12.5	[200]
Isltm1(cre/Esr1) Krc* MGI ID: 3530632	Targeted mutation, tamoxifen-inducible *cre* was inserted into exon 1 of the *Isl1* locus	Recombination in pituitary not examined, recombination cardiac progenitors	[220]

IRES, Internal ribosome entry site.

Gnrhr–internal ribosome entry site–*cre* (GRIC) mice contain an internal ribosome entry site (IRES) and *cre*-coding sequence inserted in-frame downstream of exon 3 of the *Gnrhr* gene allowing for coexpression of the *Gnrhr* gene and *cre*. GRIC mice are viable, fertile, and present in expected Mendelian ratios.[32] GRIC mice were crossed with *Rosa26-YFP* reporter mice to label cells in which CRE activity was present.[32,200] FSHB and LHB are present 99.9% of gonadotropin-containing cells labeled with YFP. YFP-positive cells are only present in 1.8% of cells lacking gonadotropins. Recombination occurs as early as e12.75.[55] Recombination also occurs ectopically in hypothalamic GnRH neurons and testis.[201]

The gonadotropin subunit, *Cga* is initially expressed in the oral ectoderm, which will form the pituitary gland. Transgenic *Cga-cre* mice have been created in which expression of *cre* is driven by 4.6 kb of proximal promoter from the mouse gene coding for the common gonadotropin α-subunit, *Cga*. In these mice, recombination occurs in pituitary primordium and all cell types of the adult anterior pituitary.[38] Ectopic expression is present in cardiac and skeletal muscle.[38] Little or no expression is detected in gonads, adrenal glands, brain, ventromedial hypothalamus, or kidneys.[38,86] Additional studies using *Cga-cre* mice indicate that initial *cre* expression occurs at approximately e9.5.[37] A similar *Cga-cre* line was developed using 4.5 kb of mouse *Cga* promoter to drive *cre* expression in the pituitary gland. Recombination occurs in pituitary, but was not tested in specific pituitary cell types. Recombination was not detected in hypothalamus.[202]

Another transgenic CRE line, *Cga-icre*, has been developed in which *CpG* content of *cre* is reduced as compared to the prokaryotic coding sequence to diminish the chance of epigenetic silencing in mammals.[203] The *cre*-coding sequence was inserted downstream of the ATG encoding the initiating methionine for *Cga*. CRE activity is detected in differentiating pituitary cells at e12.5. In adult mice CRE activity is present in gonadotropes and thyrotropes with high penetrance. Little or no activity is detected in skeletal or cardiac muscle, brain, kidneys, lungs, testis, ovary, and liver.[203]

Tetracycline-inducible *Cga-cre* transgenic mice (Bα/CreTeR) have been generated in which the –313 to +48 region of the bovine *Cga* promoter targets expression to gonadotropes, but not thyrotropes.[204] This fragment of the bovine *Cga* promoter was subcloned into the *CreTeR* vector, which contained a 1.7-kb fragment with the *cre*-coding sequence and a nuclear localization signal inserted downstream of the *TetO* sequence.[205] To characterize Bα/CreTeR mice, they were crossed with a *Rosa26 lacZ* reporter line.[206] In adult Bα/CreTeR mice, *cre* expression is induced with 7 days of doxycycline treatment. CRE activity was apparent in all gonadotropes. Other cell types were not evaluated.[205]

Other gonadotropin subunits have also been used to develop *cre* lines. *Lhb*, is expressed specifically in gonadotrope cells starting at approximately e16.5.[31] *Lhb-cre* mice were generated using 776 bp of bovine *Lhb* promoter to drive expression of a *cre*-coding sequence with an engineered consensus KOZAK sequence and a nuclear localization signal, polyadenylation signal and intron from the rabbit β-globin gene.[60] *Lhb-cre* mice crossed with a *Rosa26-GFP* reporter line or a *Rosa26-lacZ* reporter line to evaluate cell specific CRE activity reveal that recombination occurs in approximately 80% of LHB-positive cells and not in LHB-negative cells of adult mice.[60,207,208] Embryonically, one or two pituitary cells per section exhibit recombination at e16.5, while many cells have undergone recombination at e17.5 and e18.5, consistent with the expected pattern of expression for LHB.[60] CRE activity is absent from hypothalamus and adrenal gland.[60] Limited CRE activity is present in the brain and kidney.[60] Low levels of recombination occur in a few ovarian follicles and corpora lutea of female mice and in some seminiferous tubules of male mice.[60]

Expression of the gonadotropin subunit, *Fshb*, is detectable shortly before birth in the mouse.[31] Transgenic mice in which 4.7 kb of ovine *Fshb* promoter drives expression of *icre* are referred to as *Fshb-icre* mice.[209] In *Fshb-icre* mice recombination is first detected in the anterior pituitary during embryonic development. In the adult mouse, recombination occurs in 97–100% of gonadotropes and a small percentage of other pituitary cell types. Recombination was not detected in the hypothalamus, testis, or ovary.[209]

In addition to *cre* lines using hormone or receptor regulatory sequences to target recombination to gonadotropes, many transcription factor genes have been utilized for this purpose, including *Nr5a1*. *Nr5a1* codes for the orphan nuclear receptor, steroidogenic factor-1, and is highly expressed in steroidogenic tissues including gonads and adrenals, as well as nonsteroidogenic tissues, such as hypothalamus and pituitary.[30,210] Low levels of *Nr5a1* expression are initially detected in the developing pituitary gland at e13.5.[30] Transgenic mice were generated with a bacterial artificial chromosome containing 23 kb of sequence 5′ to the start site of transcription and 88 kb of the mouse *Nr5a1* gene and 3′ region. A cassette containing the *cre*-coding sequence and the bovine *Gh* polyadenylation sequence was inserted into the *Nr5a1* locus at the ATG initiator methionine.[211] *Nr5a1-cre* mice were crossed with *Rosa26-lacZ* reporter mice to characterize CRE activity.[206] *Nr5a1-cre* mice exhibit recombination in the developing pituitary gland, presumably in gonadotrope progenitor cells by e13.5.[211] CRE activity was also apparent in somatic cells of the gonads, adrenal cortex, spleen, and ventromedial hypothalamic nucleus. Additional staining was observed in the choroid plexus, hippocampus, and cerebral cortex.[211] *Nr5a1-cre* mice effectively target gonadotropes during development, as well as adrenal, gonadal, and hypothalamic tissues.

Several mouse lines have been created using *Foxl2* to direct *cre* expression.[91,212,213] Two of these lines promote recombination in response to tamoxifen.[212,213] All of these mouse models have been used to target recombination to the ovary, although to our knowledge, none has been evaluated for recombination in the pituitary gland.[91,212,213] Based on endogenous expression patterns for *Foxl2*, it is likely that these lines would target recombination to gonadotropes and thyrotropes.[27,28]

The LIM homeodomain transcription factor, *Lhx3*, is expressed in the pituitary anlage, as well as the developing spinal cord.[214–216] This expression pattern is conserved in the human pituitary and spinal cord and requires a 7.9-kb enhancer located downstream of the human *LHX3* gene.[217] *ELHX3-cre* transgenic mice were generated using this 7.9-kb enhancer from human *LHX3* and 920 bp of minimal promoter from *Hsp68* to drive expression of *cre*.[33] Recombination is detected by e14.5 in the caudomedial pituitary gland, but not in the rostral tip thyrotropes. The *LHX3* enhancer targets recombination to cells containing CGA, TSHB, FSHB, and LHB but not to cells positive for GH or ACTH when examined at birth.[33] *ELHX3-cre* mice also target recombination to V2a interneurons.[33]

Another LIM homeodomain transcription factor, *Isl1*, is expressed in developing pituitary, as well as numerous other tissues including hindlimb and heart progenitors, adrenals, motor neurons, dorsal root and cranial sensory ganglia, pancreas, and hypothalamus.[218,219] In the developing pituitary gland, *Isl1* is initially expressed in the oral ectoderm at e8.5.[96] While several *Isl1-cre* lines exists, to our knowledge none have been evaluated for recombination in the pituitary gland.[200,219,220]

5. CONCLUDING REMARKS

Evidence from mouse models has provided a glimpse into the mechanisms of gonadotrope development (Fig. 2). Signaling factors from surrounding tissue that diffuse into Rathke's pouch and signaling factors that are produced within the gland itself initiate pituitary formation by activating

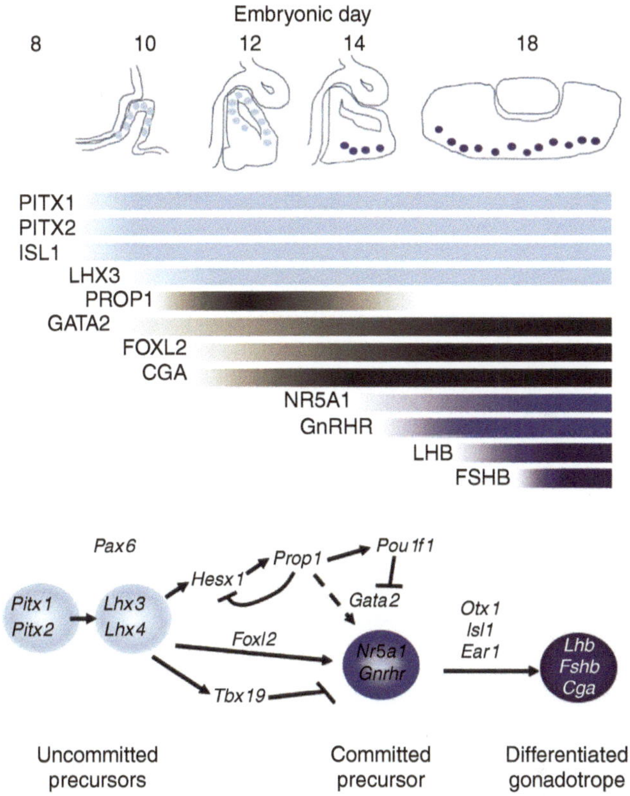

Figure 2 Uncommitted progenitor cells are rapidly dividing at e10.5 throughout Rathke's pouch. These cells express PITX1, PITX2, ISL1, LHX3, and GATA2. In the genetic hierarchy of gonadotrope development, *Pitx1* and *Pitx2* are upstream of *Lhx3* and *Lhx4*. HESX1 directly activates *Prop1* expression while PROP1 is required to extinguish expression of *Hesx1*. At e12.5, proliferating cells are located primarily around the lumen of Rathke's pouch. PROP1 directly stimulates *Pou1f1* expression and POU1F1 interacts with GATA2 via protein–protein interactions to inhibit gonadotrope differentiation. FOXL2 promotes expression of gonadotrope genes. TBX19 inhibits *Nr5a1* expression and gonadotrope differentiation. Committed gonadotrope progenitors are apparent by e14.5 and express *Nr5a1* and *Gnrhr*. OTX1, ISL1, and EGR1 stimulate expression of the gonadotropin subunits. Differentiated gonadotropes express *Lhb*, *Fshb*, and *Cga* and can be detected by e18.5.

specific transcription factors. Proper specification and differentiation of gonadotropes is dependent largely on the spatial and temporal expression patterns of these transcription factors. For example, GATA2 and POU1F1 exhibit reciprocal expression patterns and opposing effects on pituitary cell differentiation that are essential for normal gonadotrope specification.

Similar interactions occur between NR5A1 and TBX19, with ectopic expression of *Tbx19* inhibiting gonadotrope development. While a great deal of information has been gleaned from current mouse models, much remains to be learned about the mechanisms of gonadotrope development.

REFERENCES

1. Brinkley HJ. Endocrine signaling and female reproduction. *Biol Reprod.* 1981;24:22–43.
2. Gharib SD, Wierman ME, Shupnik MA, Chin WW. Molecular biology of the pituitary gonadotropins. *Endocr Rev.* 1990;11:177–199.
3. Hamernik DL, Nett TM. Gonadotropin-releasing hormone increases the amount of messenger ribonucleic acid for gonadotropins in ovariectomized ewes after hypothalamic-pituitary disconnection. *Endocrinology.* 1988;122:959–966.
4. Mason AJ, Hayflick JS, Zoeller RT, Young 3rd WS, Phillips HS, Nikolics K, Seeburg PH. A deletion truncating the gonadotropin-releasing hormone gene is responsible for hypogonadism in the hpg mouse. *Science.* 1986;234:1366–1371.
5. Clarke IJ, Cummins JT, de Kretser DM. Pituitary gland function after disconnection from direct hypothalamic influences in the sheep. *Neuroendocrinology.* 1983;36:376–384.
6. Pernasetti F, Vasilyev VV, Rosenberg SB, Bailey JS, Huang HJ, Miller WL, Mellon PL. Cell-specific transcriptional regulation of follicle-stimulating hormone-beta by activin and gonadotropin-releasing hormone in the LbetaT2 pituitary gonadotrope cell model. *Endocrinology.* 2001;142:2284–2295.
7. Vasilyev VV, Pernasetti F, Rosenberg SB, Barsoum MJ, Austin DA, Webster NJ, Mellon PL. Transcriptional activation of the ovine follicle-stimulating hormone-beta gene by gonadotropin-releasing hormone involves multiple signal transduction pathways. *Endocrinology.* 2002;143:1651–1659.
8. Roberson MS, Misra-Press A, Laurance ME, Stork PJ, Maurer RA. A role for mitogen-activated protein kinase in mediating activation of the glycoprotein hormone alpha-subunit promoter by gonadotropin-releasing hormone. *Mol Cell Biol.* 1995;15: 3531–3539.
9. Frager MS, Pieper DR, Tonetta SA, Duncan JA, Marshall JC. Pituitary gonadotropin-releasing hormone receptors. Effects of castration, steroid replacement, and the role of gonadotropin-releasing hormone in modulating receptors in the rat. *J Clin Invest.* 1981;67:615–623.
10. Bauer-Dantoin AC, Hollenberg AN, Jameson JL. Dynamic regulation of gonadotropin-releasing hormone receptor mRNA levels in the anterior pituitary gland during the rat estrous cycle. *Endocrinology.* 1993;133:1911–1914.
11. Brooks J, Taylor PL, Saunders PT, Eidne KA, Struthers WJ, McNeilly AS. Cloning and sequencing of the sheep pituitary gonadotropin-releasing hormone receptor and changes in expression of its mRNA during the estrous cycle. *Mol Cell Endocrinol.* 1993;94:R23–R27.
12. Kaiser UB, Jakubowiak A, Steinberger A, Chin WW. Regulation of rat pituitary gonadotropin-releasing hormone receptor mRNA levels in vivo and in vitro. *Endocrinology.* 1993;133:931–934.
13. Hamernik DL, Clay CM, Turzillo A, Van Kirk EA, Moss GE. Estradiol increases amounts of messenger ribonucleic acid for gonadotropin-releasing hormone receptors in sheep. *Biol Reprod.* 1995;53:179–185.
14. Ellsworth BS, White BR, Burns AT, Cherrington BD, Otis AM, Clay CM. c-Jun N-terminal kinase activation of activator protein-1 underlies homologous regulation of the gonadotropin-releasing hormone receptor gene in alphat3-1 cells. *Endocrinology.* 2003;144:839–849.

15. White BR, Duval DL, Mulvaney JM, Roberson MS, Clay CM. Homologous regulation of the gonadotropin-releasing hormone receptor gene is partially mediated by protein kinase C activation of an activator protein-1 element. *Mol Endocrinol.* 1999;13:566–577.
16. Norwitz ER, Cardona GR, Jeong KH, Chin WW. Identification and characterization of the gonadotropin-releasing hormone response elements in the mouse gonadotropin-releasing hormone receptor gene. *J Biol Chem.* 1999;274:867–880.
17. Pierce JG, Parsons TF. Glycoprotein hormones: structure and function. *Annu Rev Biochem.* 1981;50:465–495.
18. Strauss JF, Barbieri RL. *Yen & Jaffes Reproductive Endocrinology: Physiology, Pathophysiology, and Clinical Management.* 7th ed. Philadelphia, PA: Elsevier Saunders; 2013
19. Bilodeau S, Roussel-Gervais A, Drouin J. Distinct developmental roles of cell cycle inhibitors p57kip1 and p27kip2 distinguish pituitary progenitor cell cycle exit from cell cycle re-entry of differentiated cells. *Mol Cell Biol.* 2009;29(7):1895–1908.
20. Ward RD, Raetzman LT, Suh H, Stone BM, Nasonkin IO, Camper SA. Role of PROP1 in pituitary gland growth. *Mol Endocrinol.* 2005;19:698–710.
21. Ward RD, Stone BM, Raetzman LT, Camper SA. Cell proliferation and vascularization in mouse models of pituitary hormone deficiency. *Mol Endocrinol.* 2006;20:1378–1390.
22. Perez Millan MI, Brinkmeier ML, Mortensen AH, Camper SA. PROP1 triggers epithelial-mesenchymal transition-like process in pituitary stem cells. *eLIFE.* 2016;5: e14470.
23. Dasen JS, O'Connell SM, Flynn SE, Treier M, Gleiberman AS, Szeto DP, Hooshmand F, Aggarwal AK, Rosenfeld MG. Reciprocal interactions of Pit1 and GATA2 mediate signaling gradient-induced determination of pituitary cell types. *Cell.* 1999;97: 587–598.
24. Davis SW, Mortensen AH, Camper SA. Birthdating studies reshape models for pituitary gland cell specification. *Dev Biol.* 2011;352:215–227.
25. Le Tissier PR, Hodson DJ, Lafont C, Fontanaud P, Schaeffer M, Mollard P. Anterior pituitary cell networks. *Front Neuroendocrinol.* 2012;33:252–266.
26. Bonnefont X, Lacampagne A, Sanchez-Hormigo A, Fino E, Creff A, Mathieu MN, Smallwood S, Carmignac D, Fontanaud P, Travo P, Alonso G, Courtois-Coutry N, Pincus SM, Robinson IC, Mollard P. Revealing the large-scale network organization of growth hormone-secreting cells. *Proc Natl Acad Sci USA.* 2005;102:16880–16885.
27. Ellsworth BS, Egashira N, Haller JL, Butts DL, Cocquet J, Clay CM, Osamura RY, Camper SA. FOXL2 in the pituitary: molecular, genetic, and developmental analysis. *Mol Endocrinol.* 2006;20:2796–2805.
28. Treier M, Gleiberman AS, O'Connell SM, Szeto DP, McMahon JA, McMahon AP, Rosenfeld MG. Multistep signaling requirements for pituitary organogenesis in vivo. *Genes Dev.* 1998;12:1691–1704.
29. Schang AL, Granger A, Querat B, Bleux C, Cohen-Tannoudji J, Laverriere JN. GATA2-induced silencing and LIM-homeodomain protein-induced activation are mediated by a bi-functional response element in the rat GnRH receptor gene. *Mol Endocrinol.* 2013;27:74–91.
30. Ingraham HA, Lala DS, Ikeda Y, Luo X, Shen WH, Nachtigal MW, Abbud R, Nilson JH, Parker KL. The nuclear receptor steroidogenic factor 1 acts at multiple levels of the reproductive axis. *Genes Dev.* 1994;8:2302–2312.
31. Japon MA, Rubinstein M, Low MJ. In situ hybridization analysis of anterior pituitary hormone gene expression during fetal mouse development. *J Histochem Cytochem.* 1994;42:1117–1125.
32. Wen S, Schwarz JR, Niculescu D, Dinu C, Bauer CK, Hirdes W, Boehm U. Functional characterization of genetically labeled gonadotropes. *Endocrinology.* 2008;149: 2701–2711.

33. Park S, Mullen RD, Rhodes SJ. Cell-specific actions of a human LHX3 gene enhancer during pituitary and spinal cord development. *Mol Endocrinol.* 2013;27: 2013–2027.
34. McCabe MJ, Gaston-Massuet C, Tziaferi V, Gregory LC, Alatzoglou KS, Signore M, Puelles E, Gerrelli D, Farooqi IS, Raza J, Walker J, Kavanaugh SI, Tsai PS, Pitteloud N, Martinez-Barbera JP, Dattani MT. Novel FGF8 mutations associated with recessive holoprosencephaly, craniofacial defects, and hypothalamo-pituitary dysfunction. *J Clin Endocrinol Metab.* 2011;96:E1709–E1718.
35. Falardeau J, Chung WC, Beenken A, Raivio T, Plummer L, Sidis Y, Jacobson-Dickman EE, Eliseenkova AV, Ma J, Dwyer A, Quinton R, Na S, Hall JE, Huot C, Alois N, Pearce SH, Cole LW, Hughes V, Mohammadi M, Tsai P, Pitteloud N. Decreased FGF8 signaling causes deficiency of gonadotropin-releasing hormone in humans and mice. *J Clin Invest.* 2008;118:2822–2831.
36. Meyers EN, Lewandoski M, Martin GR. An Fgf8 mutant allelic series generated by Cre- and Flp-mediated recombination. *Nat Genet.* 1998;18:136–141.
37. Davis SW, Camper SA. Noggin regulates Bmp4 activity during pituitary induction. *Dev Biol.* 2007;305:145–160.
38. Cushman LJ, Burrows HL, Seasholtz AF, Lewandoski M, Muzyczka N, Camper SA. Cre-mediated recombination in the pituitary gland. *Genesis.* 2000;28:167–174.
39. Zhou X, Wang Y, Ongaro L, Boehm U, Kaartinen V, Mishina Y, Bernard DJ. Normal gonadotropin production and fertility in gonadotrope-specific Bmpr1a knockout mice. *J Endocrinol.* 2016;229:331–341.
40. Raetzman LT, Wheeler BS, Ross SA, Thomas PQ, Camper SA. Persistent expression of Notch2 delays gonadotrope differentiation. *Mol Endocrinol.* 2006;20:2898–2908.
41. Zhu X, Zhang J, Tollkuhn J, Ohsawa R, Bresnick EH, Guillemot F, Kageyama R, Rosenfeld MG. Sustained Notch signaling in progenitors is required for sequential emergence of distinct cell lineages during organogenesis. *Genes Dev.* 2006;20: 2739–2753.
42. Nantie LB, Himes AD, Getz DR, Raetzman LT. Notch signaling in postnatal pituitary expansion: proliferation, progenitors, and cell specification. *Mol Endocrinol.* 2014;28: 731–744.
43. McCright B, Lozier J, Gridley T. Generation of new Notch2 mutant alleles. *Genesis.* 2006;44:29–33.
44. Hebert JM, McConnell SK. Targeting of cre to the Foxg1 (BF-1) locus mediates loxP recombination in the telencephalon and other developing head structures. *Dev Biol.* 2000;222:296–306.
45. Raetzman LT, Cai JX, Camper SA. Hes1 is required for pituitary growth and melanotrope specification. *Dev Biol.* 2007;304:455–466.
46. Treier M, O'Connell S, Gleiberman A, Price J, Szeto DP, Burgess R, Chuang PT, McMahon AP, Rosenfeld MG. Hedgehog signaling is required for pituitary gland development. *Development.* 2001;128:377–386.
47. Park HL, Bai C, Platt KA, Matise MP, Beeghly A, Hui CC, Nakashima M, Joyner AL. Mouse Gli1 mutants are viable but have defects in SHH signaling in combination with a Gli2 mutation. *Development.* 2000;127:1593–1605.
48. Brault V, Moore R, Kutsch S, Ishibashi M, Rowitch DH, McMahon AP, Sommer L, Boussadia O, Kemler R. Inactivation of the beta-catenin gene by Wnt1-Cre-mediated deletion results in dramatic brain malformation and failure of craniofacial development. *Development.* 2001;128:1253–1264.
49. Boerboom D, Kumar V, Boyer A, Wang Y, Lambrot R, Zhou X, Rico C, Boehm U, Paquet M, Celeste C, Kimmins S, Bernard DJ. beta-catenin stabilization in gonadotropes impairs FSH synthesis in male mice in vivo. *Endocrinology.* 2015;156: 323–333.

50. Harada N, Tamai Y, Ishikawa T, Sauer B, Takaku K, Oshima M, Taketo MM. Intestinal polyposis in mice with a dominant stable mutation of the beta-catenin gene. *EMBO J.* 1999;18:5931–5942.
51. Olson LE, Tollkuhn J, Scafoglio C, Krones A, Zhang J, Ohgi KA, Wu W, Taketo MM, Kemler R, Grosschedl R, Rose D, Li X, Rosenfeld MG. Homeodomain-mediated beta-catenin-dependent switching events dictate cell-lineage determination. *Cell.* 2006;125:593–605.
52. Cattanach BM, Iddon CA, Charlton HM, Chiappa SA, Fink G. Gonadotrophin-releasing hormone deficiency in a mutant mouse with hypogonadism. *Nature.* 1977;269:338–340.
53. Charlton HM, Halpin DM, Iddon C, Rosie R, Levy G, McDowell IF, Megson A, Morris JF, Bramwell A, Speight A, Ward BJ, Broadhead J, Davey-Smith G, Fink G. The effects of daily administration of single and multiple injections of gonadotropin-releasing hormone on pituitary and gonadal function in the hypogonadal (hpg) mouse. *Endocrinology.* 1983;113:535–544.
54. Silverman AJ, Zimmerman EA, Gibson MJ, Perlow MJ, Charlton HM, Kokoris GJ, Krieger DT. Implantation of normal fetal preoptic area into hypogonadal mutant mice: temporal relationships of the growth of gonadotropin-releasing hormone neurons and the development of the pituitary/testicular axis. *Neuroscience.* 1985;16:69–84.
55. Wen S, Ai W, Alim Z, Boehm U. Embryonic gonadotropin-releasing hormone signaling is necessary for maturation of the male reproductive axis. *Proc Natl Acad Sci USA.* 2010;107:16372–16377.
56. Szeto DP, Rodriguez-Esteban C, Ryan AK, O'Connell SM, Liu F, Kioussi C, Gleiberman AS, Izpisua-Belmonte JC, Rosenfeld MG. Role of the Bicoid-related homeodomain factor Pitx1 in specifying hindlimb morphogenesis and pituitary development. *Genes Dev.* 1999;13:484–494.
57. Kioussi C, Briata P, Baek SH, Rose DW, Hamblet NS, Herman T, Ohgi KA, Lin C, Gleiberman A, Wang J, Brault V, Ruiz-Lozano P, Nguyen HD, Kemler R, Glass CK, Wynshaw-Boris A, Rosenfeld MG. Identification of a Wnt/Dvl/beta-Catenin-Pitx2 pathway mediating cell-type-specific proliferation during development. *Cell.* 2002;111:673–685.
58. Charles MA, Suh H, Hjalt TA, Drouin J, Camper SA, Gage PJ. PITX genes are required for cell survival and Lhx3 activation. *Mol Endocrinol.* 2005;19:1893–1903.
59. Suh H, Gage PJ, Drouin J, Camper SA. Pitx2 is required at multiple stages of pituitary organogenesis: pituitary primordium formation and cell specification. *Development.* 2002;129:329–337.
60. Charles MA, Mortensen AH, Potok MA, Camper SA. Pitx2 deletion in pituitary gonadotropes is compatible with gonadal development, puberty, and fertility. *Genesis.* 2008;46:507–514.
61. Sheng HZ, Zhadanov AB, Mosinger Jr B, Fujii T, Bertuzzi S, Grinberg A, Lee EJ, Huang SP, Mahon KA, Westphal H. Specification of pituitary cell lineages by the LIM homeobox gene Lhx3. *Science.* 1996;272:1004–1007.
62. Zhao Y, Morales DC, Hermesz E, Lee WK, Pfaff SL, Westphal H. Reduced expression of the LIM-homeobox gene Lhx3 impairs growth and differentiation of Rathke's pouch and increases cell apoptosis during mouse pituitary development. *Mech Dev.* 2006;123:605–613.
63. Ellsworth BS, Butts DL, Camper SA. Mechanisms underlying pituitary hypoplasia and failed cell specification in Lhx3-deficient mice. *Dev Biol.* 2008;313:118–129.
64. Sheng HZ, Moriyama K, Yamashita T, Li H, Potter SS, Mahon KA, Westphal H. Multistep control of pituitary organogenesis. *Science.* 1997;278:1809–1812.
65. Raetzman LT, Ward R, Camper SA. Lhx4 and Prop1 are required for cell survival and expansion of the pituitary primordia. *Development.* 2002;129:4229–4239.

66. Li H, Witte DP, Branford WW, Aronow BJ, Weinstein M, Kaur S, Wert S, Singh G, Schreiner CM, Whitsett JA, et al. Gsh-4 encodes a LIM-type homeodomain, is expressed in the developing central nervous system and is required for early postnatal survival. *EMBO J.* 1994;13:2876–2885.
67. Takuma N, Sheng HZ, Furuta Y, Ward JM, Sharma K, Hogan BL, Pfaff SL, Westphal H, Kimura S, Mahon KA. Formation of Rathke's pouch requires dual induction from the diencephalon. *Development.* 1998;125:4835–4840.
68. Charles MA, Saunders TL, Wood WM, Owens K, Parlow AF, Camper SA, Ridgway EC, Gordon DF. Pituitary specific Gata2 knockout: effects on gonadotrope and thyrotrope function. *Mol Endocrinol.* 2006;20(6):1366–1377.
69. Nasonkin IO, Ward RD, Raetzman LT, Seasholtz AF, Saunders TL, Gillespie PJ, Camper SA. Pituitary hypoplasia and respiratory distress syndrome in Prop1 knockout mice. *Hum Mol Genet.* 2004;13:2727–2735.
70. Vesper AH, Raetzman LT, Camper SA. Role of prophet of Pit1 (PROP1) in gonadotrope differentiation and puberty. *Endocrinology.* 2006;147:1654–1663.
71. Dattani MT, Martinez-Barbera JP, Thomas PQ, Brickman JM, Gupta R, Martensson IL, Toresson H, Fox M, Wales JK, Hindmarsh PC, Krauss S, Beddington RS, Robinson IC. Mutations in the homeobox gene HESX1/Hesx1 associated with septo-optic dysplasia in human and mouse. *Nat Genet.* 1998;19:125–133.
72. Gaston-Massuet C, Andoniadou CL, Signore M, Sajedi E, Bird S, Turner JM, Martinez-Barbera JP. Genetic interaction between the homeobox transcription factors HESX1 and SIX3 is required for normal pituitary development. *Dev Biol.* 2008;324: 322–333.
73. Acampora D, Mazan S, Tuorto F, Avantaggiato V, Tremblay JJ, Lazzaro D, di Carlo A, Mariano A, Macchia PE, Corte G, Macchia V, Drouin J, Brulet P, Simeone A. Transient dwarfism and hypogonadism in mice lacking Otx1 reveal prepubescent stage-specific control of pituitary levels of GH, FSH and LH. *Development.* 1998;125: 1229–1239.
74. Mortensen AH, Schade V, Lamonerie T, Camper SA. Deletion of OTX2 in neural ectoderm delays anterior pituitary development. *Hum Mol Genet.* 2015;24:939–953.
75. Fossat N, Chatelain G, Brun G, Lamonerie T. Temporal and spatial delineation of mouse Otx2 functions by conditional self-knockout. *EMBO Rep.* 2006;7:824–830.
76. Kioussi C, O'Connell S, St-Onge L, Treier M, Gleiberman AS, Gruss P, Rosenfeld MG. Pax6 is essential for establishing ventral-dorsal cell boundaries in pituitary gland development. *Proc Natl Acad Sci USA.* 1999;96:14378–14382.
77. Roberts RC. *Small eyes*—a new dominant eye mutation in the mouse. *Genet Res.* 1967;9:121–122.
78. St-Onge L, Sosa-Pineda B, Chowdhury K, Mansouri A, Gruss P. Pax6 is required for differentiation of glucagon-producing [alpha]-cells in mouse pancreas. *Nature.* 1997;387:406–409.
79. Bentley CA, Zidehsarai MP, Grindley JC, Parlow AF, Barth-Hall S, Roberts VJ. Pax6 is implicated in murine pituitary endocrine function. *Endocrine.* 1999;10:171–177.
80. Dellovade TL, Pfaff DW, Schwanzel-Fukuda M. The gonadotropin-releasing hormone system does not develop in Small-Eye (Sey) mouse phenotype. *Brain Res Dev Brain Res.* 1998;107:233–240.
81. Topilko P, Schneider-Maunoury S, Levi G, Trembleau A, Gourdji D, Driancourt MA, Rao CV, Charnay P. Multiple pituitary and ovarian defects in Krox-24 (NGFI-A, Egr-1)-targeted mice. *Mol Endocrinol.* 1998;12:107–122.
82. Givens ML, Rave-Harel N, Goonewardena VD, Kurotani R, Berdy SE, Swan CH, Rubenstein JL, Robert B, Mellon PL. Developmental regulation of gonadotropin-releasing hormone gene expression by the MSX and DLX homeodomain protein families. *J Biol Chem.* 2005;280:19156–19165.

83. Houzelstein D, Cohen A, Buckingham ME, Robert B. Insertional mutation of the mouse Msx1 homeobox gene by an nlacZ reporter gene. *Mech Dev.* 1997;65: 123–133.
84. Pulichino AM, Vallette-Kasic S, Tsai JP, Couture C, Gauthier Y, Drouin J. Tpit determines alternate fates during pituitary cell differentiation. *Genes Dev.* 2003;17: 738–747.
85. Ikeda Y, Luo X, Abbud R, Nilson JH, Parker KL. The nuclear receptor steroidogenic factor 1 is essential for the formation of the ventromedial hypothalamic nucleus. *Mol Endocrinol.* 1995;9:478–486.
86. Zhao L, Bakke M, Krimkevich Y, Cushman LJ, Parlow AF, Camper SA, Parker KL. Steroidogenic factor 1 (SF1) is essential for pituitary gonadotrope function. *Development.* 2001;128:147–154.
87. Yu RN, Ito M, Saunders TL, Camper SA, Jameson JL. Role of Ahch in gonadal development and gametogenesis. *Nat Genet.* 1998;20:353–357.
88. Justice NJ, Blount AL, Pelosi E, Schlessinger D, Vale W, Bilezikjian LM. Impaired FSH {beta} expression in the pituitaries of Foxl2 mutant animals. *Mol Endocrinol.* 2011;25:1404–1415.
89. Uda M, Ottolenghi C, Crisponi L, Garcia JE, Deiana M, Kimber W, Forabosco A, Cao A, Schlessinger D, Pilia G. Foxl2 disruption causes mouse ovarian failure by pervasive blockage of follicle development. *Hum Mol Genet.* 2004;13:1171–1181.
90. Tran S, Zhou X, Lafleur C, Calderon MJ, Ellsworth BS, Kimmins S, Boehm U, Treier M, Boerboom D, Bernard DJ. Impaired fertility and FSH synthesis in gonadotrope-specific Foxl2 knockout mice. *MolEndocrinol.* 2013;27:407–421.
91. Uhlenhaut NH, Jakob S, Anlag K, Eisenberger T, Sekido R, Kress J, Treier AC, Klugmann C, Klasen C, Holter NI, Riethmacher D, Schutz G, Cooney AJ, Lovell-Badge R, Treier M. Somatic sex reprogramming of adult ovaries to testes by FOXL2 ablation. *Cell.* 2009;139:1130–1142.
92. Gumbel JH, Patterson EM, Owusu SA, Kabat BE, Jung DO, Simmons J, Hopkins T, Ellsworth BS. The forkhead transcription factor, Foxd1, is necessary for pituitary luteinizing hormone expression in mice. *PLoS One.* 2012;7:e52156.
93. Wang J, Scully K, Zhu X, Cai L, Zhang J, Prefontaine GG, Krones A, Ohgi KA, Zhu P, Garcia-Bassets I, Liu F, Taylor H, Lozach J, Jayes FL, Korach KS, Glass CK, Fu XD, Rosenfeld MG. Opposing LSD1 complexes function in developmental gene activation and repression programmes. *Nature.* 2007;446:882–887.
94. Wang H, Graham I, Hastings R, Gunewardena S, Brinkmeier ML, Conn PM, Camper SA, Kumar TR. Gonadotrope-specific deletion of Dicer results in severely suppressed gonadotropins and fertility defects. *J Biol Chem.* 2015;290:2699–2714.
95. Harfe BD, McManus MT, Mansfield JH, Hornstein E, Tabin CJ. The RNaseIII enzyme Dicer is required for morphogenesis but not patterning of the vertebrate limb. *Proc Natl Acad Sci USA.* 2005;102:10898–10903.
96. Ericson J, Norlin S, Jessell TM, Edlund T. Integrated FGF and BMP signaling controls the progression of progenitor cell differentiation and the emergence of pattern in the embryonic anterior pituitary. *Development.* 1998;125:1005–1015.
97. Ohuchi H, Hori Y, Yamasaki M, Harada H, Sekine K, Kato S, Itoh N. FGF10 acts as a major ligand for FGF receptor 2 IIIb in mouse multi-organ development. *Biochem Biophys Res Commun.* 2000;277:643–649.
98. Kelberman D, Rizzoti K, Lovell-Badge R, Robinson IC, Dattani MT. Genetic regulation of pituitary gland development in human and mouse. *Endocr Rev.* 2009;30: 790–829.
99. Zhang W, Johnson JI, Tsai PS. Fgf8-deficient mice compensate for reduced GnRH neuronal population and exhibit normal testicular function. *Front Endocrinol.* 2015;6:151.

100. Lee KB, Khivansara V, Santos MM, Lamba P, Yuen T, Sealfon SC, Bernard DJ. Bone morphogenetic protein 2 and activin A synergistically stimulate follicle-stimulating hormone beta subunit transcription. *J Mol Endocrinol.* 2007;38:315–330.
101. Ho CC, Bernard DJ. Bone morphogenetic protein 2 signals via BMPR1A to regulate murine follicle-stimulating hormone beta subunit transcription. *Biol Reprod.* 2009;81:133–141.
102. Mishina Y, Hanks MC, Miura S, Tallquist MD, Behringer RR. Generation of Bmpr/Alk3 conditional knockout mice. *Genesis.* 2002;32:69–72.
103. Mishina Y, Suzuki A, Ueno N, Behringer RR. Bmpr encodes a type I bone morphogenetic protein receptor that is essential for gastrulation during mouse embryogenesis. *Genes Dev.* 1995;9:3027–3037.
104. Balistreri CR, Madonna R, Melino G, Caruso C. The emerging role of Notch pathway in ageing: focus on the related mechanisms in age-related diseases. *Ageing Res Rev.* 2016;29:50–65.
105. Raetzman LT, Ross SA, Cook S, Dunwoodie SL, Camper SA, Thomas PQ. Developmental regulation of Notch signaling genes in the embryonic pituitary: Prop1 deficiency affects Notch2 expression. *Dev Biol.* 2004;265:329–340.
106. Chuang PT, McMahon AP. Vertebrate Hedgehog signaling modulated by induction of a Hedgehog-binding protein. *Nature.* 1999;397:617–621.
107. Salisbury TB, Binder AK, Grammer JC, Nilson JH. Maximal activity of the luteinizing hormone beta-subunit gene requires beta-catenin. *Mol Endocrinol.* 2007;21:963–971.
108. Davis SW, Mortensen AH, Keisler JL, Zacharias AL, Gage PJ, Yamamura K, Camper SA. beta-catenin is required in the neural crest and mesencephalon for pituitary gland organogenesis. *BMC Dev Biol.* 2016;16:16.
109. Ciccone NA, Kaiser UB. The biology of gonadotroph regulation. *Curr Opin Endocrinol Diabetes Obes.* 2009;16:321–327.
110. Thompson IR, Kaiser UB. GnRH pulse frequency-dependent differential regulation of LH and FSH gene expression. *Mol Cell Endocrinol.* 2014;385:28–35.
111. Clayton RN, Catt KJ. Gonadotropin-releasing hormone receptors: characterization, physiological regulation, and relationship to reproductive function. *Endocr Rev.* 1981;2:186–209.
112. Desjardins C. Endocrine signaling and male reproduction. *Biol Reprod.* 1981;24:1–21.
113. Holley SJ, Hall SB, Mellon PL. Complementary expression of IGF-II and IGFBP-5 during anterior pituitary development. *Dev Biol.* 2002;244:319–328.
114. Duval DL, Nelson SE, Clay CM. A binding site for steroidogenic factor-1 is part of a complex enhancer that mediates expression of the murine gonadotropin-releasing hormone receptor gene. *Biol Reprod.* 1997;56:160–168.
115. Duval DL, Nelson SE, Clay CM. The tripartite basal enhancer of the gonadotropin-releasing hormone (GnRH) receptor gene promoter regulates cell-specific expression through a novel GnRH receptor activating sequence. *Mol Endocrinol.* 1997;11:1814–1821.
116. Ellsworth BS, Burns AT, Escudero KW, Duval DL, Nelson SE, Clay CM. The gonadotropin releasing hormone (GnRH) receptor activating sequence (GRAS) is a composite regulatory element that interacts with multiple classes of transcription factors including Smads, AP-1 and a forkhead DNA binding protein. *Mol Cell Endocrinol.* 2003;206:93–111.
117. Kam KY, Jeong KH, Norwitz ER, Jorgensen EM, Kaiser UB. Oct-1 and nuclear factor Y bind to the SURG-1 element to direct basal and gonadotropin-releasing hormone (GnRH)-stimulated mouse GnRH receptor gene transcription. *Mol Endocrinol.* 2005;19:148–162.
118. Lents CA, Farmerie TA, Cherrington BD, Clay CM. Multiple core homeodomain binding motifs differentially contribute to transcriptional activity of the murine gonadotropin-releasing hormone receptor gene promoter. *Endocrine.* 2009;35:356–364.

119. Rosenfeld MG, Briata P, Dasen J, Gleiberman AS, Kioussi C, Lin C, O'Connell SM, Ryan A, Szeto DP, Treier M. Multistep signaling and transcriptional requirements for pituitary organogenesis in vivo. *Recent Prog Horm Res.* 2000;55:1–13.
120. Gage PJ, Camper SA. Pituitary homeobox 2, a novel member of the bicoid-related family of homeobox genes, is a potential regulator of anterior structure formation. *Hum Mol Genet.* 1997;6:457–464.
121. Semina EV, Reiter R, Leysens NJ, Alward WL, Small KW, Datson NA, Siegel-Bartelt J, Bierke-Nelson D, Bitoun P, Zabel BU, Carey JC, Murray JC. Cloning and characterization of a novel bicoid-related homeobox transcription factor gene, RIEG, involved in Rieger syndrome. *Nat Genet.* 1996;14:392–399.
122. Tremblay JJ, Lanctot C, Drouin J. The pan-pituitary activator of transcription, Ptx1 (pituitary homeobox 1), acts in synergy with SF-1 and Pit1 and is an upstream regulator of the Lim-homeodomain gene Lim3/Lhx3. *Mol Endocrinol.* 1998;12:428–441.
123. Lamonerie T, Tremblay JJ, Lanctot C, Therrien M, Gauthier Y, Drouin J. Ptx1, a bicoid-related homeo box transcription factor involved in transcription of the pro-opiomelanocortin gene. *Genes Dev.* 1996;10:1284–1295.
124. Tremblay JJ, Marcil A, Gauthier Y, Drouin J. Ptx1 regulates SF-1 activity by an interaction that mimics the role of the ligand-binding domain. *EMBO J.* 1999;18:3431–3441.
125. Tremblay JJ, Drouin J. Egr-1 is a downstream effector of GnRH and synergizes by direct interaction with Ptx1 and SF-1 to enhance luteinizing hormone beta gene transcription. *Mol Cell Biol.* 1999;19:2567–2576.
126. Jorgenson RJ, Levin LS, Cross HE, Yoder F, Kelly TE. The Rieger syndrome. *Am J Med Genet.* 1978;2:307–318.
127. Gage PJ, Suh H, Camper SA. Dosage requirement of Pitx2 for development of multiple organs. *Development.* 1999;126:4643–4651.
128. Kitamura K, Miura H, Miyagawa-Tomita S, Yanazawa M, Katoh-Fukui Y, Suzuki R, Ohuchi H, Suehiro A, Motegi Y, Nakahara Y, Kondo S, Yokoyama M. Mouse Pitx2 deficiency leads to anomalies of the ventral body wall, heart, extra- and periocular mesoderm and right pulmonary isomerism. *Development.* 1999;126:5749–5758.
129. Lin CR, Kioussi C, O'Connell S, Briata P, Szeto D, Liu F, Izpisua-Belmonte JC, Rosenfeld MG. Pitx2 regulates lung asymmetry, cardiac positioning and pituitary and tooth morphogenesis. *Nature.* 1999;401:279–282.
130. Lu MF, Pressman C, Dyer R, Johnson RL, Martin JF. Function of Rieger syndrome gene in left-right asymmetry and craniofacial development. *Nature.* 1999;401:276–278.
131. Tremblay JJ, Goodyer CG, Drouin J. Transcriptional properties of Ptx1 and Ptx2 isoforms. *Neuroendocrinology.* 2000;71:277–286.
132. Suszko MI, Lo DJ, Suh H, Camper SA, Woodruff TK. Regulation of the rat follicle-stimulating hormone beta-subunit promoter by activin. *Mol Endocrinol.* 2003;17:318–332.
133. Hunter CS, Rhodes SJ. LIM-homeodomain genes in mammalian development and human disease. *Mol Biol Rep.* 2005;32:67–77.
134. Pfaeffle RW, Savage JJ, Hunter CS, Palme C, Ahlmann M, Kumar P, Bellone J, Schoenau E, Korsch E, Bramswig JH, Stobbe HM, Blum WF, Rhodes SJ. Four novel mutations of the LHX3 gene cause combined pituitary hormone deficiencies with or without limited neck rotation. *J Clin Endocrinol Metab.* 2007;92:1909–1919.
135. Pfaff SL, Mendelsohn M, Stewart CL, Edlund T, Jessell TM. Requirement for LIM homeobox gene Isl1 in motor neuron generation reveals a motor neuron-dependent step in interneuron differentiation. *Cell.* 1996;84:309–320.
136. Granger A, Bleux C, Kottler ML, Rhodes SJ, Counis R, Laverriere JN. The LIM-homeodomain proteins Isl-1 and Lhx3 act with steroidogenic factor 1 to enhance gonadotrope-specific activity of the gonadotropin-releasing hormone receptor gene promoter. *Mol Endocrinol.* 2006;20:2093–2108.

137. Camper SA, Saunders TL, Katz RW, Reeves RH. The Pit-1 transcription factor gene is a candidate for the murine Snell dwarf mutation. *Genomics.* 1990;8:586–590.
138. Gordon DF, Lewis SR, Haugen BR, James RA, McDermott MT, Wood WM, Ridgway EC. Pit-1 and GATA-2 interact and functionally cooperate to activate the thyrotropin beta-subunit promoter. *J Biol Chem.* 1997;272:24339–24347.
139. Cushman LJ, Watkins-Chow DE, Brinkmeier ML, Raetzman LT, Radak AL, Lloyd RV, Camper SA. Persistent Prop1 expression delays gonadotrope differentiation and enhances pituitary tumor susceptibility. *Hum Mol Genet.* 2001;10:1141–1153.
140. Davis SW, Keisler JL, Perez-Millan MI, Schade V, Camper SA. All hormone-producing cell types of the pituitary intermediate and anterior lobes derive from Prop1-expressing progenitors. *Endocrinology.* 2016;157:1385–1396.
141. Muzumdar MD, Tasic B, Miyamichi K, Li L, Luo L. A global double-fluorescent Cre reporter mouse. *Genesis.* 2007;45:593–605.
142. Hermesz E, Mackem S, Mahon KA. Rpx: a novel anterior-restricted homeobox gene progressively activated in the prechordal plate, anterior neural plate and Rathke's pouch of the mouse embryo. *Development.* 1996;122:41–52.
143. Brook CG, Sanders MD, Hoare RD. Septo-optic dysplasia. *Br Med J.* 1972;3:811–813.
144. Arslanian SA, Rothfus WE, Foley Jr TP, Becker DJ. Hormonal, metabolic, and neuroradiologic abnormalities associated with septo-optic dysplasia. *Acta Endocrinol.* 1984;107:282–288.
145. Izenberg N, Rosenblum M, Parks JS. The endocrine spectrum of septo-optic dysplasia. *Clin Pediatr.* 1984;23:632–636.
146. Roessmann U. Septo-optic dysplasia (SOD) or DeMorsier syndrome. *J Clin Neuroophthalmol.* 1989;9:156–159.
147. Thomas PQ, Dattani MT, Brickman JM, McNay D, Warne G, Zacharin M, Cameron F, Hurst J, Woods K, Dunger D, Stanhope R, Forrest S, Robinson IC, Beddington RS. Heterozygous HESX1 mutations associated with isolated congenital pituitary hypoplasia and septo-optic dysplasia. *Hum Mol Genet.* 2001;10:39–45.
148. Roessmann U, Velasco ME, Small EJ, Hori A. Neuropathology of "septo-optic dysplasia" (de Morsier syndrome) with immunohistochemical studies of the hypothalamus and pituitary gland. *J Neuropathol Exp Neurol.* 1987;46:597–608.
149. Carvalho LR, Woods KS, Mendonca BB, Marcal N, Zamparini AL, Stifani S, Brickman JM, Arnhold IJ, Dattani MT. A homozygous mutation in HESX1 is associated with evolving hypopituitarism due to impaired repressor-corepressor interaction. *J Clin Invest.* 2003;112:1192–1201.
150. Carvalho LR, Brinkmeier ML, Castinetti F, Ellsworth BS, Camper SA. Corepressors TLE1 and TLE3 interact with HESX1 and PROP1. *Mol Endocrinol.* 2010;24:754–765.
151. Lagutin OV, Zhu CC, Kobayashi D, Topczewski J, Shimamura K, Puelles L, Russell HR, McKinnon PJ, Solnica-Krezel L, Oliver G. Six3 repression of Wnt signaling in the anterior neuroectoderm is essential for vertebrate forebrain development. *Genes Dev.* 2003;17:368–379.
152. Coss D, Thackray VG, Deng CX, Mellon PL. Activin regulates luteinizing hormone beta-subunit gene expression through Smad-binding and homeobox elements. *Mol Endocrinol.* 2005;19:2610–2623.
153. Acampora D, Mazan S, Avantaggiato V, Barone P, Tuorto F, Lallemand Y, Brulet P, Simeone A. Epilepsy and brain abnormalities in mice lacking the Otx1 gene. *Nat Genet.* 1996;14:218–222.
154. Acampora D, Mazan S, Lallemand Y, Avantaggiato V, Maury M, Simeone A, Brulet P. Forebrain and midbrain regions are deleted in Otx2-/- mutants due to a defective anterior neuroectoderm specification during gastrulation. *Development.* 1995;121:3279–3290.

155. Liu W, Selever J, Lu MF, Martin JF. Genetic dissection of Pitx2 in craniofacial development uncovers new functions in branchial arch morphogenesis, late aspects of tooth morphogenesis and cell migration. *Development.* 2003;130:6375–6385.
156. Kaufman MH, Chang HH, Shaw JP. Craniofacial abnormalities in homozygous Small eye (Sey/Sey) embryos and newborn mice. *J Anat.* 1995;186(pt 3):607–617.
157. Chandra A, Lan S, Zhu J, Siclari VA, Qin L. Epidermal growth factor receptor (EGFR) signaling promotes proliferation and survival in osteoprogenitors by increasing early growth response 2 (EGR2) expression. *J Biol Chem.* 2013;288:20488–20498.
158. Tian J, Li Z, Han Y, Jiang T, Song X, Jiang G. The progress of early growth response factor 1 and leukemia. *Intractable Rare Dis Res.* 2016;5:76–82.
159. McMahon AP, Champion JE, McMahon JA, Sukhatme VP. Developmental expression of the putative transcription factor Egr-1 suggests that Egr-1 and c-fos are coregulated in some tissues. *Development.* 1990;108:281–287.
160. Watson MA, Milbrandt J. Expression of the nerve growth factor-regulated NGFI-A and NGFI-B genes in the developing rat. *Development.* 1990;110:173–183.
161. Li J, Zhou J, Zhang D, Song Y, She J, Bai C. Bone marrow-derived mesenchymal stem cells enhance autophagy via PI3K/AKT signalling to reduce the severity of ischaemia/reperfusion-induced lung injury. *J Cell Mol Med.* 2015;19:2341–2351.
162. Lee SL, Sadovsky Y, Swirnoff AH, Polish JA, Goda P, Gavrilina G, Milbrandt J. Luteinizing hormone deficiency and female infertility in mice lacking the transcription factor NGFI-A (Egr-1). *Science.* 1996;273:1219–1221.
163. Le Drean Y, Liu D, Xiong F, Hew CL. Presence of distinct cis-acting elements on gonadotropin gene promoters in diverse species dictates the selective recruitment of different transcription factors by steroidogenic factor-1. *Mol Cell Endocrinol.* 1997;135:31–40.
164. Gehring WJ. Homeo boxes in the study of development. *Science.* 1987;236:1245–1252.
165. Mackenzie A, Leeming GL, Jowett AK, Ferguson MW, Sharpe PT. The homeobox gene Hox 7.1 has specific regional and temporal expression patterns during early murine craniofacial embryogenesis, especially tooth development in vivo and in vitro. *Development.* 1991;111:269–285.
166. MacKenzie A, Ferguson MW, Sharpe PT. Hox-7 expression during murine craniofacial development. *Development.* 1991;113:601–611.
167. Xie H, Cherrington BD, Meadows JD, Witham EA, Mellon PL. Msx1 homeodomain protein represses the alphaGSU and GnRH receptor genes during gonadotrope development. *Mol Endocrinol.* 2013;27:422–436.
168. Bendall AJ, Abate-Shen C. Roles for Msx and Dlx homeoproteins in vertebrate development. *Gene.* 2000;247:17–31.
169. Satokata I, Maas R. Msx1 deficient mice exhibit cleft palate and abnormalities of craniofacial and tooth development. *Nat Genet.* 1994;6:348–356.
170. Lamolet B, Pulichino AM, Lamonerie T, Gauthier Y, Brue T, Enjalbert A, Drouin J. A pituitary cell-restricted T box factor, Tpit, activates POMC transcription in cooperation with Pitx homeoproteins. *Cell.* 2001;104:849–859.
171. Ikeda Y, Shen W-H, Ingraham HA, Parker KL. Developmental expression of mouse steroidogenic factor 1, an essential regulator of the steroid hydroxylases. *Mol Endocrinol.* 1994;8:654–662.
172. Jacobs SB, Coss D, McGillivray SM, Mellon PL. Nuclear factor Y and steroidogenic factor 1 physically and functionally interact to contribute to cell-specific expression of the mouse Follicle-stimulating hormone-beta gene. *Mol Endocrinol.* 2003;17:1470–1483.
173. Park SY, Lee EJ, Emge D, Jahn CL, Jameson JL. A phenotypic spectrum of sexual development in Dax1 (Nr0b1)-deficient mice: consequence of the C57BL/6J strain on sex determination. *Biol Reprod.* 2008;79:1038–1045.

174. Achermann JC, Meeks JJ, Jameson JL. Phenotypic spectrum of mutations in DAX-1 and SF-1. *Mol Cell Endocrinol.* 2001;185:17–25.
175. Wijchers PJ, Burbach JP, Smidt MP. In control of biology: of mice, men and Foxes. *Biochem J.* 2006;397:233–246.
176. Biason-Lauber A. WNT4, RSPO1, and FOXL2 in sex development. *Semin Reprod Med.* 2012;30:387–395.
177. Blount AL, Schmidt K, Justice NJ, Vale WW, Fischer WH, Bilezikjian LM. FoxL2 and Smad3 coordinately regulate follistatin gene transcription. *J Biol Chem.* 2009;284: 7631–7645.
178. Ghochani Y, Saini JK, Mellon PL, Thackray VG. FOXL2 is involved in the synergy between activin and progestins on the follicle-stimulating hormone beta-subunit promoter. *Endocrinology.* 2012;153:2023–2033.
179. Kashimada K, Pelosi E, Chen H, Schlessinger D, Wilhelm D, Koopman P. FOXL2 and BMP2 act cooperatively to regulate follistatin gene expression during ovarian development. *Endocrinology.* 2011;152:272–280.
180. Kim S-Y, Weiss J, Tong M, Laronda MM, Lee E-J, Jameson JL. Foxl2, a forkhead transcription factor, modulates nonclassical activity of the estrogen receptor-alpha. *Endocrinology.* 2009;150:5085–5093.
181. Ottolenghi C, Omari S, Garcia-Ortiz JE, Uda M, Crisponi L, Forabosco A, Pilia G, Schlessinger D. Foxl2 is required for commitment to ovary differentiation. *Hum Mol Genet.* 2005;14:2053–2062.
182. Schmidt D, Ovitt CE, Anlag K, Fehsenfeld S, Gredsted L, Treier AC, Treier M. The murine winged-helix transcription factor Foxl2 is required for granulosa cell differentiation and ovary maintenance. *Development.* 2004;131:933–942.
183. Tran S, Lamba P, Wang Y, Bernard DJ. SMADs and FOXL2 synergistically regulate murine FSHbeta transcription via a conserved proximal promoter element. *Mol Endocrinol.* 2011;25:1170–1183.
184. Uhlenhaut NH, Treier M. Foxl2 function in ovarian development. *Mol Genet Metab.* 2006;88:225–234.
185. Veitia RA. FOXL2 versus SOX9: a lifelong "battle of the sexes". *Bioessays.* 2010;32: 375–380.
186. Crisponi L, Deiana M, Loi A, Chiappe F, Uda M, Amati P, Bisceglia L, Zelante L, Nagaraja R, Porcu S, Ristaldi MS, Marzella R, Rocchi M, Nicolino M, Lienhardt-Roussie A, Nivelon A, Verloes A, Schlessinger D, Gasparini P, Bonneau D, Cao A, Pilia G. The putative forkhead transcription factor FOXL2 is mutated in blepharophimosis/ptosis/epicanthus inversus syndrome. *Nat Genet.* 2001;27: 159–166.
187. Hatini V, Huh SO, Herzlinger D, Soares VC, Lai E. Essential role of stromal mesenchyme in kidney morphogenesis revealed by targeted disruption of Winged Helix transcription factor BF-2. *Genes Dev.* 1996;10:1467–1478.
188. Levinson RS, Batourina E, Choi C, Vorontchikhina M, Kitajewski J, Mendelsohn CL. Foxd1-dependent signals control cellularity in the renal capsule, a structure required for normal renal development. *Development.* 2005;132:529–539.
189. Jaenisch R, Bird A. Epigenetic regulation of gene expression: how the genome integrates intrinsic and environmental signals. *Nat Genet.* 2003;(33 suppl):245–254.
190. Shi Y, Lan F, Matson C, Mulligan P, Whetstine JR, Cole PA, Casero RA, Shi Y. Histone demethylation mediated by the nuclear amine oxidase homolog LSD1. *Cell.* 2004;119:941–953.
191. Metzger E, Wissmann M, Yin N, Muller JM, Schneider R, Peters AH, Gunther T, Buettner R, Schule R. LSD1 demethylates repressive histone marks to promote androgen-receptor-dependent transcription. *Nature.* 2005;437:436–439.

192. Shupnik MA. Effects of gonadotropin-releasing hormone on rat gonadotropin gene transcription in vitro: requirement for pulsatile administration for luteinizing hormone-beta gene stimulation. *Mol Endocrinol.* 1990;4:1444–1450.
193. Bouamoud N, Lerrant Y, Ribot G, Counis R. Differential stability of mRNAs coding for alpha and gonadotropin beta subunits in cultured rat pituitary cells. *Mol Cell Endocrinol.* 1992;88:143–151.
194. Weiss J, Crowley Jr WF, Jameson JL. Pulsatile gonadotropin-releasing hormone modifies polyadenylation of gonadotropin subunit messenger ribonucleic acids. *Endocrinology.* 1992;130:415–420.
195. Chedrese PJ, Kay TW, Jameson JL. Gonadotropin-releasing hormone stimulates glycoprotein hormone alpha-subunit messenger ribonucleic acid (mRNA) levels in alpha T3 cells by increasing transcription and mRNA stability. *Endocrinology.* 1994;134:2475–2481.
196. Tsutsumi M, Laws SC, Rodic V, Sealfon SC. Translational regulation of the gonadotropin-releasing hormone receptor in alpha T3-1 cells. *Endocrinology.* 1995;136: 1128–1136.
197. Tsutsumi M, Laws SC, Sealfon SC. Homologous up-regulation of the gonadotropin-releasing hormone receptor in alpha T3-1 cells is associated with unchanged receptor messenger RNA (mRNA) levels and altered mRNA activity. *Mol Endocrinol.* 1993;7: 1625–1633.
198. Kim T, Do MH, Lawson MA. Translational control of gene expression in the gonadotrope. *Mol Cell Endocrinol.* 2014;385:78–87.
199. Lodish H, Berk A, Kaiser C, Krieger M, Scott M, Bretscher A, Ploegh H. *Molecular Cell Biology.* 6th ed. New York: W. H. Freeman and company; 2008
200. Srinivas S, Watanabe T, Lin CS, William CM, Tanabe Y, Jessell TM, Costantini F. Cre reporter strains produced by targeted insertion of EYFP and ECFP into the ROSA26 locus. *BMC Dev Biol.* 2001;1:4.
201. Wen S, Gotze IN, Mai O, Schauer C, Leinders-Zufall T, Boehm U. Genetic identification of GnRH receptor neurons: a new model for studying neural circuits underlying reproductive physiology in the mouse brain. *Endocrinology.* 2011;152:1515–1526.
202. Singh SP, Wolfe A, Ng Y, DiVall SA, Buggs C, Levine JE, Wondisford FE, Radovick S. Impaired estrogen feedback and infertility in female mice with pituitary-specific deletion of estrogen receptor alpha (ESR1). *Biol Reprod.* 2009;81:488–496.
203. Perez-Millan MI, Zeidler MG, Saunders TL, Camper SA, Davis SW. Efficient, specific, developmentally appropriate cre-mediated recombination in anterior pituitary gonadotropes and thyrotropes. *Genesis.* 2013;51:785–792.
204. Kendall SK, Saunders TL, Jin L, Lloyd RV, Glode LM, Nett TM, Keri RA, Nilson JH, Camper SA. Targeted ablation of pituitary gonadotropes in transgenic mice. *Mol Endocrinol.* 1991;5:2025–2036.
205. Naik K, Pittman It. Wolfe A, Miller RS, Radovick S, Wondisford FE. A novel technique for temporally regulated cell type-specific Cre expression and recombination in the pituitary gonadotroph. *J Mol Endocrinol.* 2006;37:63–69.
206. Soriano P. Generalized lacZ expression with the ROSA26 Cre reporter strain. *Nat Genet.* 1999;21:70–71.
207. Friedrich G, Soriano P. Promoter traps in embryonic stem cells: a genetic screen to identify and mutate developmental genes in mice. *Genes Dev.* 1991;5:1513–1523.
208. Zambrowicz BP, Imamoto A, Fiering S, Herzenberg LA, Kerr WG, Soriano P. Disruption of overlapping transcripts in the ROSA beta geo 26 gene trap strain leads to widespread expression of beta-galactosidase in mouse embryos and hematopoietic cells. *Proc Natl Acad Sci USA.* 1997;94:3789–3794.
209. Wang H, Hastings R, Miller WL, Kumar TR. Fshb-iCre mice are efficient and specific Cre deleters for the gonadotrope lineage. *Mol Cell Endocrinol.* 2016;419:124–138.

210. Luo X, Ikeda Y, Parker KL. A cell-specific nuclear receptor is essential for adrenal and gonadal development and sexual differentiation. *Cell.* 1994;77:481–490.
211. Bingham NC, Verma-Kurvari S, Parada LF, Parker KL. Development of a steroidogenic factor 1/Cre transgenic mouse line. *Genesis.* 2006;44:419–424.
212. Liu C, Peng J, Matzuk MM, Yao HH. Lineage specification of ovarian theca cells requires multicellular interactions via oocyte and granulosa cells. *Nat Commun.* 2015;6:6934.
213. Zheng W, Zhang H, Gorre N, Risal S, Shen Y, Liu K. Two classes of ovarian primordial follicles exhibit distinct developmental dynamics and physiological functions. *Hum Mol Genet.* 2014;23:920–928.
214. Bach I, Rhodes SJ, Pearse RV, Heinzel T, Gloss B, Scully KM, Sawchenko PE, Rosenfeld MG. P-Lim, a LIM homeodomain factor, is expressed during pituitary organ and cell commitment and synergizes with Pit-1. *Proc Natl Acad Sci USA.* 1995;92: 2720–2724.
215. Seidah NG, Barale JC, Marcinkiewicz M, Mattei MG, Day R, Chretien M. The mouse homeoprotein mLIM-3 is expressed early in cells derived from the neuroepithelium and persists in adult pituitary. *DNA Cell Biol* 1994;13(2):1163–1180
216. Zhadanov AB, Bertuzzi S, Taira M, Dawid IB, Westphal H. Expression pattern of the murine LIM class homeobox gene Lhx3 in subsets of neural and neuroendocrine tissues. *Dev Dyn.* 1995;202:354–364.
217. Mullen RD, Park S, Rhodes SJ. A distal modular enhancer complex acts to control pituitary- and nervous system-specific expression of the LHX3 regulatory gene. *Mol Endocrinol.* 2012;26:308–319.
218. Tsuchida T, Ensini M, Morton SB, Baldassare M, Edlund T, Jessell TM, Pfaff SL. Topographic organization of embryonic motor neurons defined by expression of LIM homeobox genes. *Cell.* 1994;79:957–970.
219. Yang L, Cai CL, Lin L, Qyang Y, Chung C, Monteiro RM, Mummery CL, Fishman GI, Cogen A, Evans S. Isl1Cre reveals a common Bmp pathway in heart and limb development. *Development.* 2006;133:1575–1585.
220. Laugwitz KL, Moretti A, Lam J, Gruber P, Chen Y, Woodard S, Lin LZ, Cai CL, Lu MM, Reth M, Platoshyn O, Yuan JX, Evans S, Chien KR. Postnatal isl1+ cardioblasts enter fully differentiated cardiomyocyte lineages. *Nature.* 2005;433:647–653.

CHAPTER TWO

Mouse Models for the Study of Synthesis, Secretion, and Action of Pituitary Gonadotropins

T. Rajendra Kumar[1]

Edgar L. and Patricia M. Makowski Endowed Professor, Department of Obstetrics & Gynecology, University of Colorado Denver-Anschutz Medical Campus, Aurora, CO, United States

[1] Corresponding author. E-mail address: raj.kumar@ucdenver.edu

Contents

Progress in Molecular Biology and Translational Science, Volume 143
ISSN 1877-1173
http://dx.doi.org/10.1016/bs.pmbts.2016.08.006

Abstract

Gonadotropins play fundamental roles in reproduction. More than 30 years ago, *Cga* transgenic mice were generated, and more than 20 years ago, the phenotypes of *Cga* null mice were reported. Since then, numerous mouse strains have been generated and characterized to address several questions in reproductive biology involving gonadotropin synthesis, secretion, and action. More recently, extragonadal expression, and in some cases, functions of gonadotropins in nongonadal tissues have been identified. Several genomic and proteomic approaches including novel mouse genome editing tools are available now. It is anticipated that these and other emerging technologies will be useful to build an integrated network of gonadotropin signaling pathways in various tissues. Undoubtedly, research on gonadotropins will continue to provide new knowledge and allow us transcend from benchside to the bedside.

1. INTRODUCTION

Gonadotropins are the trophic glycoprotein hormones synthesized in and secreted from gonadotropes, the least abundant cell type of all the anterior pituitary gland cell types.[1,2] These are heterodimeric hormones called luteinizing hormone (LH) and follicle-stimulating hormone (FSH). They are composed of a common α-subunit that is noncovalently associated with a hormone/receptor-specific β-subunit.[1,2] LH is structurally and functionally more similar to the placental syncytiotrophoblast cell–derived chorionic gonadotropin (CG) that also shares the α-subunit. LHβ and CGβ protein sequences are quite similar. An O-glycosylated carboxy-terminal peptide (CTP) is uniquely present only in the CGβ subunit that confers a longer half-life to the hCG dimer in circulation.[3–6] The gonadotropin subunits are encoded by distinct genes localized to separate chromosomes. In humans, the *CGB* locus is organized into a multigene cluster consisting of six *CGB* genes.[1,7] The subunit assembly is a biosynthetic step that is required for dimer secretion into blood. Transcriptional[8–10] and translational[11] regulation of gonadotropins have been extensively studied. The hypothalamic decapeptide, gonadotropin-releasing hormone (GnRH), steroids, nonsteroidal peptides, activins, and inhibins, abundantly expressed in gonads are key regulators of gonadotropin secretion.[1,12–19]

Gonadotropins act in target tissues (gonads) by binding to G-protein coupled heptahelical transmembrane receptors called FSHRs and

LHRs.[20–26] In the male, FSH binds to FSHRs on Sertoli cells and regulate germ cells,[27] and both LH and CG bind to LHRs on interstitial Leydig cells to produce testosterone.[28,29] In the female, FSH binds ovarian granulosa cells to regulate aromatase and estrogen production, whereas LH binds both theca and granulosa cells. Gonadotropin action is essential for both gametogenesis and steroidogenesis.[30–32] Expression of gonadotropin receptors on extragonadal tissues has also been identified, but the in vivo functional significance of these receptors still remains controversial.[33,34] Understanding gonadotropin biology is critical because clinically, both loss- and gain-of-function mutations in gonadotropin ligands and their cognate receptors manifest in various syndromes and disorders in humans.[35–39] Moreover, in reproductive aging research[40] and artificial reproductive technology protocols, gonadotropins play prominent roles.[41–44] To develop genetic models of reproductive disorders involving gonadotropin action, several mouse strains have been generated.[45–49]

In the classical transgenic or gain-of-function approach, specific promoter sequences are used to drive the expression of reporter genes, for example, subunit-encoding transgenes, cell-visualizing agents, such as *Escherichia coli* lacZ, fluorescent fish proteins, or CRE recombinase enzyme, often randomly integrated in multiple copies.[50–54] In the loss-of-function approach, null mutations in desired loci are first created in embryonic stem (ES) cells and then the mutant allele(s) transmitted via mouse germline. As a result of this genetic modification that can be engineered as small as in a one base pair (bp) up to the megabase range, knockout mice are born without producing the protein of interest.[55–59] Multiple of loss-of-function mutants can be combined by intercrosses to discern genetic epistasis and build hierarchical interactions among genes/loci.[55–59] Similarly, a knockout mouse line can be crossed with a gain-of-function model to achieve genetic rescue. CRE-expressing transgenic lines can be used as "driver" lines to achieve cell/tissue-specific deletion of desired alleles flanked by loxP sites that are inserted by homologous recombination in the same orientation.[60–64] In this chapter, I will describe genetic models for pituitary gonadotropins.

2. α-GLYCOPROTEIN HORMONE SUBUNIT

2.1 *Cga* Transgenic Mice

During the mouse pituitary development, *Cga* gene is the first to be activated around embryonic day (E) 11.5 in the primitive pituitary primordium.[65,66]

In the fully differentiated pituitary, *Cga* is expressed in both gonadotropes and thyrotropes.[65] *Cga* mRNA is in general, abundantly expressed compared to that of *Lhb* and *Fshb* in gonadotropes. In humans, *Cga* is expressed in both the pituitary and the placenta. Nearly 3 decades ago, human *Cga*-overexpressing transgenic mice were generated.[67] A 17-kb genomic clone was microinjected and several founders were generated from which independent lines were established. *Cga* expression studies indicated that the human *Cga* transgene was appropriately expressed only in the pituitary but not the placenta. Thus, it was concluded that transacting factors regulating *Cga* expression in the pituitary but not the placenta of mice and humans were conserved.[67] The *Cga*-overexpressing mice, interestingly did not show any other overt abnormalities or reproductive deficits.[67]

2.2 *Cga* Knockout Mice

Loss-of-function mutation in glycoprotein hormone α-subunit–encoding gene was achieved by a targeted disruption strategy in ES cells and subsequently null mice ($Cga^{-/-}$) were generated.[68] As the glycoprotein hormone α-subunit is shared among LH, FSH, and thyroid-stimulating hormone (TSH), the null mice lacked these three functional heterodimers. As a result, these mutants exhibited severe hypothyroidism and dwarfism and were hypogonadal.[68] Arrested thyroid development was noticed in late gestation resulting in suppressed thyroid hormones.[68] Secondary to this loss of thyroid function and negative feedback by thyroid hormones, pituitary hypertrophy and hyperplasia with many TSHβ subunit–producing thyrotropes and eventually thyrotrope adenoma development were observed in these mice.[69] In these thyrotrope tumors, increased cell proliferation was observed. Many thyrotrope-specific transcription factors and unfolded protein response genes were upregulated. In addition, *Tceal5*, a novel tumor suppressor was also found overexpressed in these tumors.[69] Together, $Cga^{-/-}$ genetic model established the importance of thyrotropin in pituitary ontogeny and revealed that fetal gonadotropins are not required for sexual differentiation or genital development of fetuses in both sexes.[68,69]

2.3 Other *Cga* Models

In addition to the earlier described transgenic and knockout mouse models for glycoprotein hormone α-subunit, several lines of transgenic mice were developed using the promoter sequences of *Cga* gene from different species. These lines have been useful to understand the developmental regulation of the gonadotrope/thyrotrope lineages by reporter gene expression[70,71] and selective pituitary cell ablation,[71] to achieve efficient CRE-mediated recombination in thyrotropes and gonadotropes.[72–74]

3. LHβ SUBUNIT

3.1 *Lhb* Transgenic Mice

Lhb mRNA expression occurs at E13.5 and precedes that of *Fshb* during mouse pituitary development.[65,66] Indeed, LH^+/FSH^- gonadotropes were clearly apparent in the early embryonic mouse pituitary.[66] Extensive in vitro studies have documented that a combinatorial action of multiple transcription factors is required for basal- and GnRH-induced *Lhb* promoter activation.[9] Transgenic mouse models were also generated to test the in vivo functional roles of key transcription factor binding sites on the *Lhb* promoter region. For example, SF-1- and Pitx1-binding sites were mutated on *Lhb* promoter and transgenic mice with targeted expression of either the wild type or mutant promoter in each case were generated.[9] In both cases, the mutant promoters remained inactive indicating that the corresponding transcription-binding sites were physiologically important for *Lhb* gene expression.[9] Thus, the transgenic in vivo approach eliminated the need for overexpressing the transcription factor–encoding genes under in vitro culture conditions using heterologous cell lines.[9]

3.1.1 *LHβ-CTP Transgenic Mice*

To study the consequences of gain-of-function effects of LH, four different transgenic mouse models were developed.[75–78] In the first model, transgenic mice were produced that express a long-acting chimeric LH analog, called LHβ-CTP in pituitary gonadotropes using a 360-bp region of the bovine *Cga* promoter. This analog was engineered by genetically fusing the CTP-encoding nucleotide sequences from *HCGB* to that of bovine *Lhb* with the goal to enhance the LH half-life in circulation.[76] Furthermore, because LH hypersecretion has often been linked to infertility and miscarriages in women, it was reasoned that these transgenic mice would serve as a model to study these prevalent reproductive disorders.[76] Targeted expression of this *Lhb-CTP* transgene in gonadotropes resulted in elevated LH levels and transgenic females displayed infertility secondary to precocious puberty, infrequent or no ovulations, prolonged luteal phase, and excess testosterone and estradiol.[76] Particularly, the hyperandrogenemia phenotype emerged during prepubertal period.[76,79] Histopathological analyses indicated enlarged ovaries and presence of cysts and often granulosa cell tumors.[76] The granulosa cell tumors were found to be strain-dependent,[80] and the ovarian cyst formation required signaling via estrogen receptor β.[81] Further studies led to the observation that primordial follicles were rapidly depleted in the ovaries of these transgenic mice causing rapid ovarian senescence.[82] Additional reproductive tract defects were noted in these female transgenic mice and include a failure of uterine receptivity and midgestation pregnancy.[83] Several meiotic defects in oocytes, including aneuploidy, were noted and these defects could explain in part the implantation/pregnancy failure in these mice.[84] These mice also developed extragonadal defects, notably in kidneys including hydronephropathy, pyelonephritis,[76,84] and adrenal steroidogenesis via the induction of LHRs by the LH-CTP analog itself.[85] Finally, transgenic mice also developed ovary-dependent pituitary adenomas in cells of the Pit-1 lineage, that is, of lactrotrope, somatotrope, and thyrotrope origin.[86]

3.1.2 HCGβ Transgenic Mice

In the second model, transgenic mice were generated using a 36-kb cosmid clone that encompassed the six human *CGB* genes.[78] The entire cluster was engineered such that it was possible to investigate the in vivo regulation of all the human *CGB* genes in mouse placenta where *CGB* is not expressed. Although hCG is expressed in very high levels during the early human pregnancy (first trimester), *CGB* expression was low (~1% that of human *CGB*) and occurred late in gestation around E14, in the mouse placenta in transgenic mice.[78] The human placenta level of *CGB* expression matched surprisingly, only in cerebral cortex. Ectopic expression of *CGB* also occurred at low levels in the pituitary and adrenal glands of transgenic mice.[78] While *CGB* genes 5, 3, and 8 were transcriptionally active in mouse placenta in ratios similar to those observed in the human placenta, only *CGB* genes 1 and 2 were found active in the brain resulting in *CGB* transcripts encoding different proteins, unlike CGβ protein in these transgenic mice.[78] There were no other phenotypes reported in these *CGB* transgenic mice. Thus, despite some differences in the level, timing, and expression in the placental cell type,[78,87] *CGB* transgenic mice serve as a useful in vivo model to study the complex regulation of human *CGB* multigene cluster.

3.1.3 Ubiquitin c-hCGβ Mice

In the third model, transgenic mice (hCGβ$^+$) were generated in which the human *CGB* cDNA was targeted to multiple tissues using an ubiquitin

C promoter.[77] The goal was to determine the in vivo consequences of chronic exposure to elevated levels of LH/hCG. As the ubiquitin C promoter is active in multiple tissues, including pituitary gonadotropes, the encoded hCGβ subunit combined with the endogenous mouse glycoprotein hormone α-subunit and hCG dimer was consequently secreted, at low to moderate high levels.[77] Another line designated hCG$^+$, expressing high levels of hCG dimer, was also produced by intercrossing independent lines of mice expressing either α- or CGβ subunit–encoding cDNA from the ubiquitin C promoter. In this model, hCG dimer was ectopically produced from multiple tissues because they all expressed both the hCG subunits.[77,88]

Female transgenic hCGβ$^+$ mice displayed a variety of reproductive abnormalities including precocious puberty, infertility, increased ovarian steroidogenesis, and abnormal uterine development.[89] By 10–12 months of age, these mice developed advanced lactotrope-derived pituitary adenomas (prolactinomas) that were progesterone dependent.[90] As a consequence, these mice exhibited hyperprolactinemia, the characteristic prolactin (PRL) hormone hypersecretion phenotype. By 6 months of age, hyperprolactinemia also led to metabolic disturbances, including hyperinsulinemia, hypertriglyceridemia and dyslipidemia, glucose intolerance, and insulin resistance.[91] As dopamine is a natural suppressor of lactotrope growth and PRL secretion, short-term pharmacological suppression of hyperprolactinemia was achieved by treatment with a dopamine receptor-2 agonist. This treatment restored the fertility phenotype of these female transgenic mice.[92] Around 9–12 months of age, metastatic mammary adenocarcinoma was also evident in these female transgenic mice.[77] These tumors were ER- and PR-negative and showed marked upregulation of Wnt family members, Wnt7b and -5b by direct hCG action, and abnormal activation of the downstream signal transducer, β-catenin.[93] The observation that hCG induces upregulation of Wnt family members could be clinically relevant as pregnancy-induced breast cancers are common in humans.[93] In sharp contrast to females, male transgenic hCGβ$^+$ mice demonstrated a modest threefold increase in circulating levels of hCG as measured by bioactivity and a mild reproductive phenotype.[45]

Similar to transgenic hCGβ$^+$ females, hCG$^+$ transgenic females also developed precocious puberty. This phenotype was accompanied by elevated androgens, suppression of endogenous LH levels, and activated hypothalamic GnRH pulses and gene expression.[94] Blocking androgen action with flutamide, an androgen antagonist, at the onset of precocious puberty (early on from postnatal day 6 onward) reduced aromatization in the preoptic area (POA) of hypothalamus, the center that controls LH surge in the adult.[94] Transgenic hCG$^+$ females developed tumors in multiple organs, ovarian luteomas, macroadenomatous prolactinomas hypersecreting PRL, and consequently mammary gland tumors, similar to those observed in hCGβ$^+$ transgenic female mice. However, tumors in these hCG dimer–expressing mice developed much more aggressively, presumably due to stimulation by >1000-fold elevated serum hCG levels.[88] Interestingly, these hCG$^+$ female mice showed increased bone mineral density as a result of bone formation and decreased resorption that was reversed upon ovariectomy.[95] Thus, this bone phenotype was attributed in part due to steroids rather than a direct hCG action on bone, although expression analyses did indicate the presence of LHRs on osteoblasts. These LHRs did not either bind hCG ligand or elicit an induction of cAMP or phosphorylation of ERK1/2 pathways.[95]

In contrast to hCG$^+$ transgenic females, the corresponding transgenic males that expressed bioactive hCG at >2000-fold higher levels. These males did not display tumors in multiple tissues or any bone phenotypes.[95] However, hCG$^+$ transgenic males showed elevated levels of testosterone, and displayed the male sexual behavioral deficit of an inability to copulate.[96] Other age-dependent defects, such as distended seminal vesicles, blocked vas deferens, urinary bladder obstruction, and progressive tubule degeneration in testes were observed.[96] Further studies beginning postpartum day 5 up to adulthood indicated that during the prepubertal stage at the age of 10 days, prominent Leydig cell tumors developed in these transgenic males.[97] Ultrastructural and gene expression analyses confirmed that these tumors arose from the fetal Leydig cell lineage, which is distinct from that of the adult lineage.[97] Thus, hCG$^+$ transgenic male mice, to some extent, phenocopy younger boys carrying constitutively activating mutations in *LHR*.[97–100]

3.1.4 MT-hCGβ Transgenic Mice

In the fourth model, individual lines of transgenic mice that express either a 2.7-kb human *CGA* minigene or a 3.7-kb *CGB* gene, from the 1.8-kb mouse metallothinein-1 (MT) promoter, were generated.[75] As the MT promoter is ubiquitously and highly active, the subunits were ectopically and highly expressed in multiple tissues. MT-hCGα$^+$ transgenic male and female mice were normal and fertile. In contrast, MT-hCGβ$^+$ transgenic males and females were infertile.[75] While no apparent gonadal and other male reproductive tract defects were discernable in MT-hCGβ$^+$ transgenic males, MT-hCGβ$^+$ transgenic females showed several ovarian defects.[75]

These mice did not exhibit estrus cycles and hence were anovulatory. Histological analyses revealed a block in ovarian folliculogenesis with absence of antral follicles and corpora lutea (CL). Occasionally, hemorrhagic ovarian cysts and enlarged uterine horns were observed in these MT-hCGβ$^+$ transgenic females.[75]

As MT-hCGβ$^+$ transgenic mice were infertile, MT-hCGα and MT-hCGβ transgenes were coinjected into mouse embryos to generate transgenic mice that express the hCG dimer in multiple tissues.[75] The low level hCG dimer–expressing male and female mice were although initially fertile and produced normal number of viable pups, they became progressively infertile due to unknown etiology.[75] The high level hCG dimer–expressing (hCG$^+$) founder male mice were infertile and thus independent lines could not be established. These founder males exhibited aggressive male sexual behavior secondary to perhaps elevated levels of circulating testosterone and not because of estradiol, which was undetectable. They displayed enlarged seminal vesicles often fluid-filled, epididymis, and prostate (the accessory male sex glands) that are targets of androgen action. Serum LH and FSH levels were suppressed in these male mice. Histological analysis of testes showed many vacuolated tubules, pockets of germ cell aplasia resulting in Sertoli cell–only tubules, and prominent Leydig cell hyperplasia.[75]

Transgenic hCG$^+$ female mice, similar to males, were also infertile. Their serum estradiol levels were highly elevated, nearly 15 times more than those in nontransgenic control littermates consistent with presence of enlarged uterine horns.[75] These transgenic mice displayed massive hemorrhage and multiple cysts by 6–7 weeks of age. Histological analysis revealed many aberrant ovarian follicles with an enlarged thecal cell layer and proliferating stromal tissue consisting of multinucleated cells.[75] Transgenic hCG$^+$ female, but not male, mice exhibited cystic kidneys containing glomeruli with damaged architecture and enlarged urinary bladder. In these female mice, mammary or adrenal tumors were not noted.[75]

Thus, gender-specific phenotypes resulted in hCG-overexpressing mice. It is also noteworthy that different phenotypes emerged in these hCG overexpression models depending upon the site of hCG expression, that is pituitary versus ectopic tissues.

3.2 *Lhb* Knockout Mice

In naturally occurring *hypogonadal* (*hpg*) mutant mice, the GnRH peptide is absent and consequently, these mice lack both LH and FSH.[101,102] Similarly, both FSH and LH dimers are absent in *Cga* null mice owing to a deletion at the common glycoprotein α-subunit–encoding gene.[68] Thus, it is not possible to genetically test the in vivo consequences of loss of only LH or FSH in the aforementioned two mouse models. To circumvent these problems, a null mutation at the *Lhb* locus (encodes the hormone-specific LHβ subunit) was engineered by homologous recombination in ES cells, and hence deficiency of only LH was achieved in mice.[103] Loss of LH protein in pituitaries of *Lhb* null mice was confirmed by immunolabeling and western blot analysis of pituitaries, and radioimmunoassay on serum samples. Similar analyses further indicated that FSH homeostasis was unaffected in the absence of LH in *Lhb* null mice.[103]

Lhb null male mice displayed severe hypogonadism and infertility. In these mutant mice, testis size was markedly reduced and external genitalia and accessory sex glands were all hypoplastic as a result of severely suppressed serum and intratesticular testosterone.[103] Many of the mRNAs encoding the testosterone biosynthetic pathway mediators were suppressed in the mutant testis, consistent with poorly developed interstitium containing very few mature Leydig cells and elevated levels of androstenedione, an androgen precursor in serum.[103] Gene expression analyses further confirmed that fetal Leydig cells persisted and the majority of the adult type Leydig cells were absent in the testes of *Lhb* null males. Additionally, many Sertoli cell genes were aberrantly expressed in mutants. As Leydig cell–derived androgen signals via the androgen receptor (AR)–mediated signaling in Sertoli cells and this signaling is critical for spermatogenesis, defects in sperm production were noted in mutant testes.[103] Whereas spermatogonia, meiotic cells, and round spermatids were readily apparent, elongated and late-stage spermatids were absent in the mutant testes.[103] Thus, absence of LH in male mice resulted in a spermatogenesis block at the round to elongated spermatid developmental transition, a process known as spermiogenesis.

Lhb null female mice displayed small ovaries and thin uteri. Ovarian steroids (estradiol and progesterone) in serum were suppressed. Ovarian histology showed many normal primary and secondary follicles and many abnormal and apoptotic antral follicles.[103] But, prerovulatory follicles and CL were absent in ovarian sections indicating impaired estrus cycles in these null females.[103] Although the expression of many theca-specific genes was downregulated, a normal thecal cell layer was present in follicles. Expression of *Ptgs2*, a known LH-responsive marker gene for ovulation, was also significantly suppressed in null mutant ovaries.[103] As a consequence of these ovarian abnormalities in the absence of LH, *Lhb* null mice were infertile.

Although the endogenous LH ligand is absent in *Lhb* null mice, responsiveness to exogenous LH may not be lost, as expression of LHRs is maintained in gonads of these null mice. Based on this, pharmacological rescue of null mice was performed using exogenous gonadotropins.[103] Immature mutant mice at 21 days of age responded to hCG, an LH analog. In testes of hCG-injected mutant males, expression of mRNAs encoding steroidogenic markers were significantly upregulated and comparable to those levels observed in testes of control mice. Similarly, expression of ovarian marker genes went up when immature mutant females were pharmacologically rescued with hCG.[103] Immature mutant females, similar to control immature female mice, also superovulated in response to the standard eCG/hCG protocol.[103] Thus, *Lhb* null mice represent a genetic model to study the consequences of an isolated deficiency of LH in reproduction, while retaining normal LH responsiveness in target cells.

In a second set of experiments, short-term (1-week) pharmacological rescue of mutant males was preformed using testosterone pellets. Of the three Leydig cell mRNAs whose expression was analyzed in testes of treated males, that of *Hsd3b1* and *Hsd3b6* went up, whereas, *Cyp17a1* remained unchanged.[103] These rescue experiments with *Lhb* null males permitted an in vivo analysis of LH- versus testosterone-responsive genes in mouse Leydig cells. *Lhb* null males proved useful also in deciphering the role of bone-derived osteocalcin in directly regulating Leydig cell–produced testosterone.[104]

4. FSHβ SUBUNIT

Fshb is the last of the gonadotropin subunit–encoding genes activated in gonadotropes during the mouse pituitary development.[65,66] FSHβ protein expression in the mouse pituitary gonadotropes is more prominent around or after birth.[66] Expression of *Fshb*, similar to that of *Lhb*, is regulated by a combinatorial action of transcription factors, a number of locally produced growth factors within pituitary, GnRH, and steroids.[8,12,13,105–107]

4.1 *Fshb* Transgenic Mice

4.1.1 Gonadotrope-Targeted HFSHB Transgenic Mice

To study the molecular basis for the gondotrope-specific expression and hormonal regulation of the FSHβ-encoding gene in vivo, a 10-kb human *FSHB* transgene was assembled from two partially overlapping plasmid clones and microinjected into fertilized one-cell mouse embryos. Multiple transgenic founder mice were generated that harbored the transgene in a range of 1–50 copies from which independent lines were established. Three of the established lines exhibited similar phenotypic characteristics.[108] The transgene was appropriately targeted to pituitary and the encoded human FSHβ subunit was exclusively expressed in gonadotropes, as identified by immunofluorescence using a human FSHβ-specific monoclonal antibody. As free human FSHβ was undetectable in serum of transgenic mice, the transgene-derived human FSHβ presumably assembled with the endogenous mouse α-subunit within gonadotropes.[108] This resulted in the, most likely, secretion of an interspecies FSH hybrid that was detectable in serum by a rat/mouse RIA, as circulating FSH levels in transgenic mice were several times higher compared to those in control nontransgenic littermates.[108]

Despite the high-level FSH production, transgenic females did not show any overt phenotypes, whereas transgenic males showed a modest 12% increase in testis weight. Gonadectomy resulted in an increased expression and gonadectomy followed by steroid supplementation suppressed the expression of the human *FSHB* mRNA and pituitary and serum FSH levels. The suppressive effects of testosterone supplementation on postcastration rise in FSHβ-encoding mRNA occurred only in transgenic but not in normal control male mice. These studies suggested existence of species-specific differences in androgen regulation of human *FSHB* and mouse *Fshb* gene expression.[108] Thus, a 10-kb human *FSHB* transgene contains all the necessary regulatory elements for pituitary gonadotrope-specific expression and hormonal regulation in mice. Subsequent studies confirmed that this 10-kb human *FSHB* transgene was also appropriately regulated by GnRH and androgens, which suppress the expression of transgene but not of the endogenous mouse *Fshb* gene directly at the pituitary level.[109,110] Using this transgenic assay system as a starting point, extensive in vivo mapping of regulatory elements on human *FSHB* transgene were further narrowed down to −200 bp upstream on the 5′ flanking side.[111]

4.1.2 MT-hFSHβ Transgenic Mice

As pituitary-targeted expression of human *FSHB* did not result in any major phenotypes, in vivo consequences of overexpression of FSH, ectopically from multiple tissues were analyzed. To achieve a high-level expression of human FSH, first, the MT-1 promoter was used to express either a human *CGA* or *FSHB* transgene in multiple tissues of transgenic mice.[112] As $FSHB^+$ females were infertile, the fertile $FSHB^+$ males were intercrossed with CGA^+ females to generate FSH dimer–expressing mice. Mice expressing FSH at low levels developed normally without any overt phenotypes and were fertile. In sharp contrast, multiple reproductive defects were noted in mice expressing FSH at high levels (FSH^+) far exceeding than those normally seen in postmenopausal women.[112] These FSH^+ transgenic males were infertile and they displayed highly enlarged and translucent seminal vesicles secondary to elevated serum testosterone. The testes weights in FSH^+ males were comparable to those in normal control mice. Grossly normal spermatogenesis with all stages represented in healthy and intact tubules with abundant spermatozoa in the lumen.[112] While there were more sperm in the epididymis, sperm parameters in FSH^+ males were nearly identical to those in normal control mice. Leydig cells in the interstitium appeared normal. It was suggested that high levels of either serum FSH or testosterone likely would have affected male sexual behavioral characteristics and resulted in infertility of these males.[112]

Transgenic FSH^+ females were infertile and demonstrated high levels of ovarian steroids (estrogen and progesterone) and testosterone but normal levels of IGF-1 and LH in serum. The entire female reproductive tract was grossly enlarged with fluid-filled uteri and massively hemorrhagic ovaries. More than 95% of these female transgenic mice appeared cachexic and died between 6 and 9 weeks of age.[112] Ovarian histology showed foci of hemorrhagic cysts that developed as early as 2 weeks of age. By 6 weeks, there was minimal follicular architecture with arrested folliculogenesis and massive hemorrhage, and cysts became more prominent in ovaries of these mice. Very few normal and healthy follicles appeared at this stage. Very few female transgenic mice survived up to 13 weeks of age. Ovaries in these older mice contained essentially no follicles, but there were no obvious signs of any tumor, either.[112] Thus, gain-of-function phenotypes, mostly resembling human ovarian hyperstimulation/polycystic ovarian syndromes were apparent in these transgenic mice. These in vivo experiments suggest that elevated levels of FSH do not directly cause ovarian tumors in FSH^+ transgenic female mice.

Another genetic model was developed to study the effects of FSH alone in the absence of LH on testis development. In this model, first, human FSH subunit–encoding cDNAs were engineered in tandem and expressed from a rat insulin II promoter.[113,114] The resulting FSH dimer–expressing mice were eventually crossed onto and maintained on the *hpg* genetic background. This genetic approach provided a model to test the effects of human FSH in the absence of LH.[113,114] Biologically active FSH dimers were detectable in serum of transgenic mice at levels ranging between 0.1 and 25 IU/L, typically noted in fertile humans. Only male transgenic mice with serum FSH levels > 1 IU/L, partially rescued the *hpg* testis phenotype. Testis weights were significantly increased as a result of increased number of Sertoli cells in transgenic males. Histological analysis of the testes showed that germ cell maturation progressed beyond meiosis, but spermiogenesis was incomplete with only minimal initiation and presence of round spermatids and very few elongated spermatids.[113,114] This testis phenotype closely resembled that seen in *Lhb* null male mice.[103] Supplementation with testosterone had a synergistic effect along with FSH and resulted in a dramatic enhancement of postmeiotic germ cell development.[113,115] Thus, this model allowed delineating the independent roles of testosterone and FSH in the absence of LH, on distinct aspects of Sertoli/germ cell development during spermatogenesis.

4.1.3 Igf1-hFSHβ Transgenic Mice

In transgenic female mice expressing FSH on *hpg* background, ovaries were markedly enlarged with increased primordial follicle reserve,[116] follicular recruitment and development to advanced antral follicles occured, and serum inhibin A and B levels were elevated.[114,116] Transgenic female mice alone (not on *hpg* background) displayed an accelerated ovarian aging that resulted in premature ovarian failure by the age of 23 weeks and beyond.[117] An unexpected extragonadal phenotype, namely increased bone mass was also observed in transgenic mice expressing FSH on *hpg* background. The increase in bone mass was attributed to an indirect effect of FSH on osteoblasts and likely mediated via ovary-derived inhibin A or testosterone in females and males, respectively.[118]

The aforementioned genetics models for FSH yielded valuable information with regard to biological roles of FSH in vivo. Other studies reported generation and characterization of several lines of transgenic mice using *Fshb* regulatory/promoter regions from different species. These mice also provided useful genetic tools to study transcriptional regulation of *Fshb* gene,[119–121] gonadotrope development, tumorigenesis,[122] and gonadotrope cell ablation,[123] and to achieve gonadotrope-specific expression of CRE recombinase to facilitate recombination events in this lineage.[66]

4.2 *Fshb* Knockout Mice

To investigate the in vivo physiological consequences of FSH and model human reproductive disorders involving the FSH signaling pathway, a null mutation was engineered at the *Fshb* locus in mouse ES cells.[124] *Fshb* gene consists of three exons and the engineered mutation deleted exons 1, 2, and nearly most of the exon 3. From the targeted mutant ES cells, eventually *Fshb* null mice were generated in a normal Mendelian ratio.[124] RNA and protein expression analyses indicated that FSH was absent in these null mice.

Fshb null males were fertile when tested beginning at 6 weeks of age. The testis size was reduced in these null male mice with reduced tubule volume[124] as a result of reduced number of Sertoli cells quantified by stereology.[125] However, the net number of Leydig cells per testis and serum LH and testosterone levels were unaffected in mutant males indicating that LH signaling pathway was intact.[124,125] This also reflected in grossly normal seminal vesicles and epidymis, the targets of androgen action. Despite a reduction in testis size and the number of Sertoli cells, all stages of spermatogenesis appeared grossly normal. Analysis of sperm parameters revealed the epididymal sperm number was reduced by 75% and the number of motile sperm decreased by 40%, but sperm viability was not affected.[124] Finnish men harboring mutations in *FSHR* showed similar phenotypes as did *Fshb* null males.[124,126]

Fshb null females, in contrast to null males, were infertile and demonstrated small ovaries, hypoplastic uteri, and did not exhibit estrus cycles. Ovarian histology showed that primordial, primary, and early secondary follicles were normal, but there was a preantral stage block in folliculogenesis and CL were absent in mutant females.[124] Most of the follicles in which an apparent antrum was beginning to form appeared atretic. Serum estradiol levels were undetectable and there was a 50% reduction in serum progesterone. Serum LH levels increased with age in *Fshb* null females, a phenotype that was noted as gender specific and not seen in null males. This could be due to differences in the suppressive effects of steroids (negative feedback) between the two sexes.[124] Thus, *Fshb* null female mice phenocopy patients with human ovarian dysgenesis syndrome, a familial genetic disorder in which FSH signaling via FSHR is affected.[124,127,128] Immature female *Fshb* null mice superovulated in a pharmacological rescue experiment, indicating that FSH responsiveness was not lost in these mice.[124]

Further characterization of *Fshb* null female mice indicated that in the absence of FSH, "neurite"-like transzonal projections (TZPs) manifested and oriented granulosa cells toward the oocyte, where apical centrosomes positioned at the site of granulosa cell–zona contact.[129] FSH treatment resulted in a "withdrawal" of these TZPs and initiated changes in oocyte transcriptional activity and meiotic competence. The TZP retraction could be one mechanism by which FSH modulates somatic–germ cell communication during ovarian follicle development.[129] Further, recent studies indicated that FSH also regulates granulosa cell–oocyte communication via an increase in connexin-mediated gap junctions and cadherin-mediated cell attachment.[130] FSH was shown to play two key roles in late folliculogenesis, by regulating EGFRs on cumulus granulosa cells and meiotic maturation.[131] The consequences of loss of FSH on female reproductive tract were also analyzed during aging. In aged *Fshb* null females, enlarged and noncontractile uterine masses were detected. These mice also demonstrated ovarian growth and hypertrophy of the interstitial tissue, ovarian cysts, and epithelial and tubal inclusions, somewhat resembling human serous ovarian adenocarcinomas.[132]

Contrary to the FSH induction of bone phenotypes identified in insulin II promoter–hFSH$^+$ transgenic mice, increased bone mass was observed in the absence of FSH in *Fshb* null female mice.[133–135] Assessment of bone parameters by μCT measurements, and serum analyses indicated that loss of FSH contributed to the increase. Expression analyses by western blotting, qPCR, immunolocalization, *Fshr* reporter assays, and cell sorting using a variety of bone marrow–derived cells, primary human osteoclasts, and cell lines indicated that FSH receptors were present on osteoclasts but not osteoblasts.[134,135] In osteoclasts, FSH bound to FSHRs, coupled through $G_{i2\alpha}$ pathway, phosphorylated ERK, p38, JNK, and AKT downstream and stimulated *c-fos*.[134,135] As elevated serum FSH levels correlate much better than the decline in estrogen during peri–postmenopausal transition in women, this extragonadal FSH action on bone opened up a new avenue of research on osteoporosis (further discussed in "Beyond Reproduction: Pituitary Hormone Actions on Bone"). Several others have also reported expression of extragonadal FSHRs on uterine cells,[136] placenta,[136] adipose tissue,[137] and tumor blood vessels.[138,139] Whether these extragonadal FSHRs are functionally relevant in a physiological setting needs to be carefully evaluated in the future.[33]

4.3 Genetic Rescue of *Fshb* Null Mice

Following the characterization of *Fshb* null mice, a genetic rescue strategy was developed to introduce human *FSHB* transgenes onto *Fshb* null background, and to determine whether an interspecies FSH hybrid is biologically active in vivo and rescues FSH deficiency.[140] In this strategy, on an *Fshb* null background, the pituitary-targeted human *FSHB* transgene (type I rescue, gonadotrope-specific) and ectopically expressed *MT-CGA/FSHB* transgenes (type II rescue, with low-level FSH expression from multiple tissues) were separately introduced.[140] In the type I rescue males, the testis size, tubule volume, sperm number, and sperm motility were all restored to those in normal control males. The type I rescue female mice resumed normal estrus cycles, and exhibited all stages of normal ovarian folliculogenesis. These mice were fertile and delivered normal size litters.[140] The type II rescue males, similar to type I rescue, were fully rescued and demonstrated normal testis and sperm parameters. However, only 30% of (3 out of 10) type II rescue females were fertile, became pregnant, and exhibited normal parturition.[140] Thus, genetic rescue experiments confirmed that an interspecies FSH hybrid that resulted from gonadotrope-targeted expression of a human *FSHB* transgene rescued *Fshb* deficiency. Both the gene regulation and function of this human *FSHB* transgene in the mouse pituitary were indistinguishable from the endogenous mouse *Fshb* gene. Finally, *Fshb* genetic rescue studies illustrated that mutations could be introduced into specific loci in the mouse genome by a gene targeting approach, without global perturbations. This is because restoring the FSH protein expression, completely rescued FSHβ deficiency created by a null mutation at the *Fshb* locus.[140]

4.4 FSH Rerouted Mice

The applicability of *Fshb* null mice as a genetic platform to test the in vivo bioactivity of FSH or FSH analogs was illustrated by two recent models. In the first case, a transgene encoding a human FSHβ–LHβ–C′ heptapeptide chimera was engineered and introduced onto *Fshb* null background. As the endogenous mouse FSHβ was absent in these rescue mice, the intragonadotrope behavior and gonadal function of only this transgene-derived chimera was assessed.[141] The presence of the LHβ–C′ heptapeptide redirected the mutant FSHβ containing FSH dimer into the LH secretory pathway where it was packaged into dense core granules, responded to acute GnRH stimulation, and likely resulted in a pulsatile mode of FSH secretion, like LH. The rerouted FSH caused a dramatic increase in the number of ovulations per estrus cycle, enhanced ovarian function for a prolonged period of time by preventing atresia, and altered the ovarian gene responses.[141] These studies revealed that the C′ heptapeptide of LHβ is a gonadotrope-sorting determinant in vivo and the target cells, that is, ovarian granulosa cells can interpret how the hormone signal inputs (constitutive versus pulsatile FSH) are presented.

4.5 FSH Glycosylation Mutant Mice

Microheterogeneity due to terminal sugars has been known to influence FSH bio/immunological activity, serum half-life, receptor-binding kinetics, and signal transduction.[142–146] On the contrary, the existence of macroheterogeneous FSH glycoforms has been discovered[147–153] but the in vivo roles of these glycoforms are unknown. In the second example, the in vivo functional significance of N-glycosylation sites on human FSHβ (Asn^7 and Asn^{24}, where sugar chains are normally added en bloc) was tested in the context of macroheterogeneity, that is, whether the entire sugar chain at one N-glycosylation site was present or absent.[154] A human *FSHB* transgene encoding a double N-glycosylation site mutant FSHβ ($Asn^{7\Delta 24\Delta}$) subunit was targeted to gonadotropes on an *Fshb* null background. Although the mutant FSHβ protein was localized to LH^+ gonadotropes and readily detected by Western blot analysis under denaturing conditions, no or very low levels of FSH dimers were detected in pituitary extracts when subjected to nondenaturing SDS-PAGE followed by immunoblotting.[154] FSH secretion into medium was also undetectable when pituitary fragments from these mice were cultured short term. Serum FSH was similarly undetectable in these mice.[154] Both male and female *Fshb* null mice carrying the transgene encoding the double glycosylation site mutant FSHβ did not show any rescue.[154] In these mice, the gonad size did not increase, and gene expression in gonads was similar to that in *Fshb* null mice.[154] These in vivo studies indicated that the double N-glycosylation site mutant FSHβ subunit failed to combine with the mouse α-subunit. Even when assembled in low levels, FSH dimer containing this mutant FSHβ subunit was secretion incompetent and failed to rescue *Fshb* null mice. Thus, at least one N-glycosylation site must be present on the FSHβ subunit for efficient FSH dimer assembly and its secretion in vivo.[154]

5. TRANSFORMING GROWTH FACTOR-β SUPERFAMILY MEMBERS AND GONADOTROPINS

The transforming growth factor (TGF)-β superfamily is the largest family of secreted growth factor proteins.[155–160] They regulate a wide variety of growth, developmental, and physiological processes. In addition to these functions, some members are highly expressed in the hypothalamus, pituitary, and gonads and control reproduction.[16,161–163] Typically, the ligands bind and form a trimeric complex with the cognate receptor type II and type I. The phosphorylated type I receptor then recruits a common signal transducer protein, SMAD4 along with a receptor-specific SMAD. This phosphorylated SMAD complex then translocates into the nucleus and transcriptionally regulates target genes.[155,157,159] Here, mouse models for some of these TGF-β family members and signaling proteins are described.

5.1 Inhibins and Activins

Inhibin was a peptide originally purified from gonadal extracts and named for its role in inhibiting pituitary FSH release.[12] It is highly expressed in gonads, particularly Sertoli cells in the testis and granulosa cells in the ovary. A more widespread expression in many tissues was later identified.[12,164] It is a heterodimer of an α-subunit and either a βA or βB subunit, and was accordingly named Inhibin A or B, respectively.[12,14,164] Homo- or heterodimers of the β-subunits are called activins, and three activins were identified, namely, activin A (βA homodimer), B (βB homodimer), and AB (βA and βB heterodimer).[12,14,16,164–167] Although two other activin subunits, βC and βE, and the resulting homodimers, activin C and E were identified, their in vivo roles are not clear.[168,169] Activins bind and signal through activing type IIA or type IIB receptors.[12,170,171] Although activin type IIA is expressed throughout the reproductive axis,[164,172] the expression of type IIB is more restricted to nonreproductive tissues, such as heart, digestive, and skeletal systems.[16,173]

5.2 *Inha* Knockout Mice

To define the in vivo roles of inhibin, *Inha* knockout mice were generated, in which the α-subunit gene was mutated. As a result, both inhibin A and B were absent, but activin subunits were still expressed in these *Inha* null mice.[174] Although inhibin is expressed in multiple tissues, *Inha* knockout mice developed hemorrhagic gonadal sex cord stromal tumors with 100% penetrance as early as 4 weeks of age, and their serum FSH, activins, and estradiol levels were elevated. These mice developed cachexia (the wasting syndrome) and liver apoptosis secondary to elevated activins, which signal via the type IIA receptor.[164,174,175] As a result of the tumor burden and wasting syndrome, nearly 90% of males died by 12 weeks and 95% of females died by 22 weeks of age.[164,174,175] Gonadectomy resulted in the development of adrenal cortical tumors that were often metastatic.[164,174,175] These genetic studies identified that inhibin is a secreted tumor suppressor factor for gonads and adrenals.

5.3 *Inha*/Gonadotropin Double Knockout Mice

To delineate the role of gonadotropins in gonadal sex cord tumor development in *Inha* null mice, three double mutant strains were developed. First, in *Inha/hpg* double mutant mice, as a result of loss of both gonadotropins, FSH and LH, gonadal and adrenal tumors did not develop. These mice did not develop the wasting syndrome either and survived for more than 1 year.[176] Second, *Inha/Fshb* mice were developed to directly test whether FSH contributes to gonadal tumor formation and/or progression in the absence of Inhibin.[112] In contrast to *Inha* male mice, 90% of which died by 12 weeks of age, only 30% of *Inha/Fshb* double mutant males died by 1 year of age. These double mutants had minimal cachexia and displayed slow-growing and less hemorrhagic gonadal tumors.[112] Similarly, 70% of double mutant female mice lived past 17 weeks, at which age, >95% of *Inha* female mice died. The milder cachexia phenotypes in double mutant *Inha/Fshb* mice were correlated with reduced levels of serum activins, estradiol, and aromatase in gonadal tumors.[112] Thus, FSH was identified as an important trophic modifier factor for gonadal tumors in *Inha* null mice. Third, *Inha/Lhb* double mutant mice were generated to test the role of LH in gonadal tumor formation in the absence of inhibin.[177] These studies indicated that LH is not required for tumor formation but in the absence of LH, gonadal tumors are significantly delayed as a result of reduction in serum FSH, estradiol, cyclin D2, and cell cycle inhibitors in tumors in these double mutant mice.[177] Ovarian transplantation experiments definitively established that inhibin was the primary tumor suppressor and various modifier factors influence the gonadal tumorigenesis in *Inha* null mice.[164] Although the search for a bona fide inhibin receptor continued for a long time, only a binding protein, InhBP/p120 (also known as betaglycan) was biochemically identified and further characterized.[178–180] However, mutant mice lacking this binding protein were normal and did not show any overt phenotypes indicating that physiologically, signaling via this binding protein is not essential for inhibin bioactivity in vivo or for reproductive function.[178]

5.4 Knockout Mice for Activin Ligands

Activins promote FSH release in pituitary cell cultures.[12,105,164,171] Activin receptor type IIA is evolutionarily conserved and expressed in multiple tissues including the hypothalamus, pituitary gonadotropes, and gonads.[172,181] Knockout mouse strains lacking *Inhba*,[163] *Inhbb*,[182] and *Acvr2a*[172] genes were separately developed. *Inhba* null mice died perinatally because they lacked whiskers and failed to suckle, displayed cleft palate, and other craniofacial defects.[163] Hence, it was not possible to study FSH regulation in these null mice. *Inhbb* null mice were born with open eyelids and developed other eye defects. Serum FSH levels were normal and comparable to those in wild type and heterozygous control mice.[182] These null males were normal and fertile, but null females exhibited increased gestation time and defects in nursing ability due to impaired oxytocin-mediated milk ejection that caused perinatal lethality of their offspring.[182]

5.5 Activin Receptor2A Knockout Mice

Although a minor proportion die embryonically due to lack of incisors, mandible, and other craniofacial defects, the majority of *Acvr2a* null mice exhibited defects in FSH homeostasis suggesting signaling via this receptor is important in pituitary gonadotropes for FSH production.[172] Secondary to suppressed FSH levels, testis and sperm parameters were decreased in null males, a proportion of which also exhibited delayed puberty and were infertile.[172,183] Combined loss of FSH and ACVRIIA in *Fshb/Acvr2a* double mutant males further confirmed that locally produced activins also signal via ACVR2A to regulate Sertoli cell number and spermatogenesis.[183] Finally, the majority of the infertile *Acvr2a* null males exhibited male sexual behavioral defects, independent of hypothalamic GnRH, LH, and testosterone.[181] These defects were due to a reduction in the activity of neuronal nitric oxide synthase exclusively in the POA of the hypothalamus, where ACVR2A is normally highly expressed in control mice.[181] *Acvr2a* null females were infertile and displayed thin uteri and small ovaries. Histologically, ovaries showed pronounced follicular atresia and rarely CL, which suggested impaired estrus cycles in these null female mice.[172] Thus, these in vivo genetic experiments revealed that different phenotypes manifest in mice lacking activins or ACVR2A. Importantly, consistent with the known biochemical data, these experiments indicated promiscuity among TGF-β ligands and receptors and that activins did not signal via ACVR2A during embryogenesis.[172]

5.6 Knockout Mice for Smads

Several mouse mutants were produced lacking signal transduction proteins downstream of activins and bone morphogenetic proteins (BMPs). SMAD2 and SMAD3 proteins partner with SMAD4 and transduce activin ligand actions via type II and type I receptors.[12,16] Gonadotrope-specific deletion of *Smad2* or *Smad3* encoding a DNA binding–deficient mutant or a combination of both did not result in major changes in FSH.[184] In contrast, mutant mice with a gonadotrope-specific deletion of *Smad4* showed a significant reduction in *Fshb* mRNA and circulating FSH levels and manifested subfertility in males and females.[185] Deletion of activin receptor 1C–encoding gene *Acvr1c* or *Alk7* in mice resulted in defects all along the reproductive axis only in females, consistent with its normal broad expression pattern in the female reproductive axis.[186,187] In the hypothalamus of mutant females, expression of *Npy* mRNA was reduced. The number of *Npy*-expressing neurons was also reduced and the NPY^+ $AgRP^+$ innervation in the POA was lost.[186] Mutant females exhibited delayed puberty, abnormal estrus cycles, suppressed gonadotropins, premature ovarian failure accompanied by degenerating oocytes, and impaired exogenous gonadotropin responsiveness.[186] In contrast, deletion of *Bmpr1a* in gonadotropes did not affect FSH homeostasis, although BMPs signal through BMPR1A in vitro.[188]

5.7 *Gdf9* Knockout Mice

Growth differentiation factor-9 (GDF-9) was originally discovered as an oocyte-specific secreted growth factor that influences granulosa function in various stages of folliculogenesis.[189] Loss-of-function mutation in *Gdf9* gene resulted in hypergonadotropic hypogonadism and ovarian folliculogenesis arrest.[189] Recently, in cell culture models GDF-9 was shown to modulate GnRH pulse frequency sensitivity of *Fshb* transcription via an ALK5–Smad2/3 pathway.[190] Whether this GDF-9 regulated pathway affects *Fshb* transcription in vivo in response to GnRH pulses remains to be tested.

5.8 Follistatin Transgenic Mice

Follistatin (FS) is a binding protein, abundantly expressed in gonads, pituitary, and other tissues.[14,19,164,191] It is also expressed in several tissues similar to activins and inhibins. In many tissues, it binds and presents activin, as well as acts as an antagonist and neutralizes activin signaling.[14,19,164,191] Transgenic mice overexpessing FS were generated. Similar to overexpression of FSH and hCG, MT-1 promoter was used to express FS in multiple tissues.[192] All of the five transgenic founders were initially fertile; three of the lines with the highest level of FS expression showed defects in testis, ovary, and hair. In the testes of these lines, variable Leydig cell hyperplasia and arrested spermatogenesis with degenerating tubules leading to infertility was noted.[192] In two lines, ovarian folliculogenesis defects at various stages were observed resembling those observed in mutants lacking other TGF-β superfamily members. Although, FS is known to suppress FSH levels normally, via activin signaling, FSH levels were suppressed in only one line with the most widespread expression of FS.[192] Thus, overexpression of FS in multiple tissues leads to defects in gonads and pituitary.

5.9 Follistatin Knockout Mice

Knockout mice were generated to study the in vivo consequences of loss of function of FS. These mice died soon after birth due to defects in multiple tissues, including whiskers, skin, and others.[193] This precluded analysis of gonad development and function in the adult. In one study, to test the role of FS in testis development and function, fetal testes were obtained from FS null mice at E18, and transplanted to the external ear of immunocompromised castrated male mice.[194] The grafted testes when analyzed after 7–8 weeks of transplantation, showed full spermatogenesis. This study indicated that secreted, but not locally produced FS from the host testis, could support the development of testes obtained from FS null mice. In a second study, to determine the postnatal effects of FS on ovarian development, granulosa cell–specific deletion of *Fst* was achieved with an *Amhr2Cre*^{KI} driver line.[195] These mice showed reduced numbers of ovarian follicles, ovulation and fertilization defects, elevated levels of serum gonadotropins, reduced levels of testosterone, and in some cases, infertility. Thus, loss of ovarian granulosa cell–derived FS in mice led to a premature ovarian failure phenotype, often seen in women.[195]

6. OTHER MOUSE MODELS

Similar to the characterization of gain- and loss-of-function mouse models for gonadotropins described earlier, genetic models for their cognate receptors, mimicking mutations often seen at the clinic were also developed. For example, gain-of-function mouse models with constitutively active LHRs[196,197] and FSHRs[198,199] have been developed. Loss-of-function mouse models lacking *Lhr*[200,201] and *Fshr*[202,203] mostly phenocopy the corresponding ligand mutants, and in addition, show some unique phenotypes.[47,204] Mice with gonadotrope-specific deletion of genes encoding transcriptional regulators,[185,205–209] or other factors[210,211] and steroid hormone receptors[212–215] that directly regulate gonadotropin synthesis and secretion, have also been developed and extensively characterized. All of these models serve as valuable in vivo tools to study gonadotropin biology and physiology in a whole-animal context.

7. CONCLUSIONS AND FUTURE DIRECTIONS

Built on a solid foundation of classical physiology, biochemistry, and cell biology, the gonadotropin field of research has become tremendously advanced. Mouse molecular genetics approaches have provided greater insights into gonadotropin biosynthesis, secretion, and action. Individual mutant mouse strains have been combined to answer key questions and identify in vivo genetic interactions along the reproductive axis. Novel genome editing techniques should prove useful in rapidly generating mouse mutants to unravel the physiological roles of gonadotropin signaling components in reproductive physiology.[216–220] Such methods should also allow us answer unambiguously, the in vivo functional significance of extragonadal roles of gonadotropins. Mouse models for different FSH glycoforms could be developed and the in vivo consequences systematically tested.[154,221] Knowledge gained through mouse models should also help refine gonadotropin-based fertility treatments, as mutant models missing endogenous gonadotropins could serve as potential preclinical animal models. The effects of excessive stimulation by gonadotropin treatment and the epigenetic consequences have been realized.[43,222] These should be further investigated in the future with the aid of mouse models. Finally, although the efficacy of gonadotropin-based vaccines to block fertility is still controversial, gonadotropin peptidomimetics and small molecule drugs to block gonadotropin action[223–226] could be tested in mouse models with exaggerated gonadotropin action.

ACKNOWLEDGMENTS

I thank Prof. P. Michael Conn for his comments on this chapter. Research in my laboratory on mouse models for gonadotropins has been supported by the National Institutes of Health (NIA, NCI, NIGMS, and NICHD); The Hall family Foundation, Kansas City, MO, USA; and the Edgar L. and Patricia M. Makowski Endowment at the University of Colorado Denver, Anschutz Medical Campus, Aurora, CO, USA.

REFERENCES

1. Bousfield GR, Jia L, Ward DN, Gonadotropins: chemistry and biosynthesis. 3rd ed. Neill JD, ed. *Knobil and Neills Physiology of Reproduction.* vol. 1. New York: Elsevier Press; 2006:1581–1634.
2. Pierce JG, Parsons TF. Glycoprotein hormones: structure and function. *Annu Rev Biochem.* 1981;50:465–495.
3. Boime I, Ben-Menahem D. Glycoprotein hormone structure-function and analog design. *Recent Prog Horm Res.* 1999;54:271–288.
4. Furuhashi M, Shikone T, Fares FA, Sugahara T, Hsueh AJ, Boime I. Fusing the carboxy-terminal peptide of the chorionic gonadotropin (CG) beta-subunit to the common alpha-subunit: retention of O-linked glycosylation and enhanced in vivo bioactivity of chimeric human CG. *Mol Endocrinol.* 1995;9:54–63.
5. Muyan M, Boime I. The carboxyl-terminal region is a determinant for the intracellular behavior of the chorionic gonadotropin beta subunit: effects on the processing of the asn-linked oligosaccharides. *Mol Endocrinol.* 1998;12:766–772.
6. Muyan M, Furuhashi M, Sugahara T, Boime I. The carboxy-terminal region of the beta-subunits of luteinizing hormone and chorionic gonadotropin differentially influence secretion and assembly of the heterodimers. *Mol Endocrinol.* 1996;10:1678–1687.
7. Policastro PF, Daniels-McQueen S, Carle G, Boime I. A map of the hCG beta-LH beta gene cluster. *J Biol Chem.* 1986;261:5907–5916.
8. Bernard DJ, Fortin J, Wang Y, Lamba P. Mechanisms of FSH synthesis: what we know, what we don't, and why you should care. *Fertil Steril.* 2010;93:2465–2485.
9. Jorgensen JS, Quirk CC, Nilson JH. Multiple and overlapping combinatorial codes orchestrate hormonal responsiveness and dictate cell-specific expression of the genes encoding luteinizing hormone. *Endocr Rev.* 2004;25:521–542.
10. Nilson JH, Bokar JA, Clay CM, Farmerie TA, Fenstermaker RA, Hamernik DL, Keri RA. Different combinations of regulatory elements may explain why placenta-specific expression of the glycoprotein hormone alpha-subunit gene occurs only in primates and horses. *Biol Reprod.* 1991;44:231–237.
11. Kim T, Do MH, Lawson MA. Translational control of gene expression in the gonadotrope. *Mol Cell Endocrinol.* 2014;385:78–87.
12. Bilezikjian LM, Blount AL, Donaldson CJ, Vale WW. Pituitary actions of ligands of the TGF-beta family: activins and inhibins. *Reproduction.* 2006;132:207–215.
13. Ciccone NA, Kaiser UB. The biology of gonadotroph regulation. *Curr Opin Endocrinol Diabetes Obes.* 2009;16:321–327.
14. de Kretser DM, Hedger MP, Loveland KL, Phillips DJ. Inhibins, activins and follistatin in reproduction. *Hum Reprod Update.* 2002;8:529–541.
15. Mahesh VB. The dynamic interaction between steroids and gonadotropins in the mammalian ovulatory cycle. *Neurosci Biobehav Rev.* 1985;9:245–260.
16. Namwanje M, Brown CW. Activins and inhibins: roles in development, physiology, and disease. *Cold Spring Harb Perspect Biol.* 2016;8(7):a021881.
17. Thompson IR, Kaiser UB. GnRH pulse frequency-dependent differential regulation of LH and FSH gene expression. *Mol Cell Endocrinol.* 2014;385:28–35.
18. Tong S, Wallace EM, Burger HG. Inhibins and activins: clinical advances in reproductive medicine. *Clin Endocrinol.* 2003;58:115–127.
19. Welt C, Sidis Y, Keutmann H, Schneyer A. Activins, inhibins, and follistatins: from endocrinology to signaling. A paradigm for the new millennium. *Exp Biol Med.* 2002;227:724–752.
20. Bogerd J. Ligand-selective determinants in gonadotropin receptors. *Mol Cell Endocrinol.* 2007;260–262:144–152.
21. Fowler PA, Huhtaniemi IT. The ovarian gonadotropin receptors in health and disease. *Rev Endocr Metab Disord.* 2002;3:55–63.

22. Menon KM, Menon B. Structure, function and regulation of gonadotropin receptors—a perspective. *Mol Cell Endocrinol.* 2012;356:88–97.
23. Ulloa-Aguirre A, Reiter E, Bousfield G, Dias JA, Huhtaniemi I. Constitutive activity in gonadotropin receptors. *Adv Pharmacol.* 2014;70:37–80.
24. Segaloff DL. Diseases associated with mutations of the human lutropin receptor. *Prog Mol Biol Transl Sci.* 2009;89:97–114.
25. Segaloff DL. Regulatory processes governing the cell surface expression of LH and FSH receptors. *Subcell Biochem.* 2012;63:113–129.
26. Tao YX, Segaloff DL. Follicle stimulating hormone receptor mutations and reproductive disorders. *Prog Mol Biol Transl Sci.* 2009;89:115–131.
27. Griswold MD, Heckert L, Linder C. The molecular biology of the FSH receptor. *J Steroid Biochem Mol Biol.* 1995;53:215–218.
28. Ascoli M, Segaloff DL. On the structure of the luteinizing hormone/chorionic gonadotropin receptor. *Endocr Rev.* 1989;10:27–44.
29. Latronico AC, Segaloff DL. Naturally occurring mutations of the luteinizing-hormone receptor: lessons learned about reproductive physiology and G protein-coupled receptors. *Am J Hum Genet.* 1999;65:949–958.
30. Richards JS, Jahnsen T, Hedin L, Lifka J, Ratoosh S, Durica JM, Goldring NB. Ovarian follicular development: from physiology to molecular biology. *Recent Prog Horm Res.* 1987;43:231–276.
31. Richards JS, Pangas SA. New insights into ovarian function. *Handb Exp Pharmacol.* 2010;3–27.
32. Richards JS, Russell DL, Ochsner S, Hsieh M, Doyle KH, Falender AE, Lo YK, Sharma SC. Novel signaling pathways that control ovarian follicular development, ovulation, and luteinization. *Recent Prog Horm Res.* 2002;57:195–220.
33. Kumar TR. Extragonadal FSH receptor: is it real? *Biol Reprod.* 2014;91:99.
34. Pakarainen T, Ahtiainen P, Zhang FP, Rulli S, Poutanen M, Huhtaniemi I. Extragonadal LH/hCG action—not yet time to rewrite textbooks. *Mol Cell Endocrinol.* 2007;269:9–16.
35. Bhagavath B, Layman LC. The genetics of hypogonadotropic hypogonadism. *Semin Reprod Med.* 2007;25:272–286.
36. Huhtaniemi I. Activating and inactivating hormone receptor mutations. *Horm Res.* 2000;53:9–16.
37. Huhtaniemi I. The Parkes lecture. Mutations of gonadotrophin and gonadotrophin receptor genes: what do they teach us about reproductive physiology? *J Reprod Fertil.* 2000;119:173–186.
38. Huhtaniemi IT. LH and FSH receptor mutations and their effects on puberty. *Horm Res.* 2002;57(suppl 2):35–38.
39. Siegel ET, Kim HG, Nishimoto HK, Layman LC. The molecular basis of impaired follicle-stimulating hormone action: evidence from human mutations and mouse models. *Reprod Sci.* 2013;20:211–233.
40. Perheentupa A, Huhtaniemi I. Aging of the human ovary and testis. *Mol Cell Endocrinol.* 2009;299:2–13.
41. Castillo JC, Humaidan P, Bernabeu R. Pharmaceutical options for triggering of final oocyte maturation in ART. *Biomed Res Int.* 2014;2014:580171.
42. Ezcurra D, Humaidan P. A review of luteinising hormone and human chorionic gonadotropin when used in assisted reproductive technology. *Reprod Biol Endocrinol.* 2014;12:95.
43. Pacchiarotti A, Selman H, Valeri C, Napoletano S, Sbracia M, Antonini G, Biagiotti G, Pacchiarotti A. Ovarian stimulation protocol in IVF: an up-to-date review of the literature. *Curr Pharm Biotechnol.* 2016;17:303–315.
44. Pouwer AW, Farquhar C, Kremer JA. Long-acting FSH versus daily FSH for women undergoing assisted reproduction. *Cochrane Database Syst Rev.* 2012;13(6). CD009577.

45. Jonas KC, Oduwole OO, Peltoketo H, Rulli SB, Huhtaniemi IT. Mouse models of altered gonadotrophin action: insight into male reproductive disorders. *Reproduction.* 2014;148:R63–R70.
46. Peltoketo H, Rivero-Muller A, Ahtiainen P, Poutanen M, Huhtaniemi I. Consequences of genetic manipulations of gonadotrophins and gonadotrophin receptors in mice. *Ann Endocrinol.* 2010;71:170–176.
47. Ratner LD, Rulli SB, Huhtaniemi IT. Genetically modified mouse models addressing gonadotropin function. *Reprod Biol.* 2014;14:9–15.
48. Kumar TR. What have we learned about gonadotropin function from gonadotropin subunit and receptor knockout mice? *Reproduction.* 2005;130:293–302.
49. Kumar TR. Mouse models for gonadotropins: a 15-year saga. *Mol Cell Endocrinol.* 2007;260–262:249–254.
50. Brinster RL. Stem cells and transgenic mice in the study of development. *Int J Dev Biol.* 1993;37:89–99.
51. Brinster RL, Palmiter RD. Introduction of genes into the germ line of animals. *Harvey Lect.* 1984;80:1–38.
52. Palmiter RD, Brinster RL. Transgenic mice. *Cell.* 1985;41:343–345.
53. Palmiter RD, Brinster RL. Germ-line transformation of mice. *Annu Rev Genet.* 1986;20:465–499.
54. Furuta Y, Behringer RR. Recent innovations in tissue-specific gene modifications in the mouse. *Birth Defects Res C Embryo Today.* 2005;75:43–57.
55. Ayadi A, Birling MC, Bottomley J, Bussell J, Fuchs H, Fray M, Gailus-Durner V, Greenaway S, Houghton R, Karp N, Leblanc S, Lengger C, et al. Mouse large-scale phenotyping initiatives: overview of the European Mouse Disease Clinic (EUMODIC) and of the Wellcome Trust Sanger Institute Mouse Genetics Project. *Mamm Genome.* 2012;23:600–610.
56. Mills AA, Bradley A. From mouse to man: generating megabase chromosome rearrangements. *Trends Genet.* 2001;17:331–339.
57. van der Weyden L, Adams DJ, Bradley A. Tools for targeted manipulation of the mouse genome. *Physiol Genomics.* 2002;11:133–164.
58. van der Weyden L, Bradley A. Mouse chromosome engineering for modeling human disease. *Annu Rev Genomics Hum Genet.* 2006;7:247–276.
59. van der Weyden L, Shaw-Smith C, Bradley A. Chromosome engineering in ES cells. *Methods Mol Biol.* 2009;530:49–77.
60. Belizario JE, Akamini P, Wolf P, Strauss B, Xavier-Neto J. New routes for transgenesis of the mouse. *J Appl Genet.* 2012;53:295–315.
61. Bouabe H, Okkenhaug K. Gene targeting in mice: a review. *Methods Mol Biol.* 2013;1064:315–336.
62. Deng CX. Conditional knockout mouse models of cancer. *Cold Spring Harb Protoc.* 2014;2014:1217–1233.
63. Lewandoski M. Analysis of mouse development with conditional mutagenesis. *Handb Exp Pharmacol.* 2007;235–262.
64. Schmidt-Supprian M, Rajewsky K. Vagaries of conditional gene targeting. *Nat Immunol.* 2007;8:665–668.
65. Japon MA, Rubinstein M, Low MJ. In situ hybridization analysis of anterior pituitary hormone gene expression during fetal mouse development. *J Histochem Cytochem.* 1994;42:1117–1125.
66. Wang H, Hastings R, Miller WL, Kumar TR. Fshb-iCre mice are efficient and specific Cre deleters for the gonadotrope lineage. *Mol Cell Endocrinol.* 2016;419:124–138.
67. Fox N, Solter D. Expression and regulation of the pituitary- and placenta-specific human glycoprotein hormone alpha-subunit gene is restricted to the pituitary in transgenic mice. *Mol Cell Biol.* 1988;8:5470–5476.

68. Kendall SK, Samuelson LC, Saunders TL, Wood RI, Camper SA. Targeted disruption of the pituitary glycoprotein hormone alpha-subunit produces hypogonadal and hypothyroid mice. *Genes Dev.* 1995;9:2007–2019.
69. Gergics P, Christian HC, Choo MS, Ajmal A, Camper SA. Gene expression in mouse thyrotrope adenoma: transcription elongation factor stimulates proliferation. *Endocrinology.* 2016;en-2016.
70. Kendall SK, Gordon DF, Birkmeier TS, Petrey D, Sarapura VD, O'Shea KS, Wood WM, Lloyd RV, Ridgway EC, Camper SA. Enhancer-mediated high level expression of mouse pituitary glycoprotein hormone alpha-subunit transgene in thyrotropes, gonadotropes, and developing pituitary gland. *Mol Endocrinol.* 1994;8:1420–1433.
71. Kendall SK, Saunders TL, Jin L, Lloyd RV, Glode LM, Nett TM, Keri RA, Nilson JH, Camper SA. Targeted ablation of pituitary gonadotropes in transgenic mice. *Mol Endocrinol.* 1991;5:2025–2036.
72. Charles MA, Saunders TL, Wood WM, Owens K, Parlow AF, Camper SA, Ridgway EC, Gordon DF. Pituitary-specific Gata2 knockout: effects on gonadotrope and thyrotrope function. *Mol Endocrinol.* 2006;20:1366–1377.
73. Cushman LJ, Burrows HL, Seasholtz AF, Lewandoski M, Muzyczka N, Camper SA. Cre-mediated recombination in the pituitary gland. *Genesis.* 2000;28:167–174.
74. Perez-Millan MI, Zeidler MG, Saunders TL, Camper SA, Davis SW. Efficient, specific, developmentally appropriate cre-mediated recombination in anterior pituitary gonadotropes and thyrotropes. *Genesis.* 2013;51:785–792.
75. Matzuk MM, DeMayo FJ, Hadsell LA, Kumar TR. Overexpression of human chorionic gonadotropin causes multiple reproductive defects in transgenic mice. *Biol Reprod.* 2003;69:338–346.
76. Risma KA, Clay CM, Nett TM, Wagner T, Yun J, Nilson JH. Targeted overexpression of luteinizing hormone in transgenic mice leads to infertility, polycystic ovaries, and ovarian tumors. *Proc Natl Acad Sci USA.* 1995;92:1322–1326.
77. Rulli SB, Kuorelahti A, Karaer O, Pelliniemi LJ, Poutanen M, Huhtaniemi I. Reproductive disturbances, pituitary lactotrope adenomas, and mammary gland tumors in transgenic female mice producing high levels of human chorionic gonadotropin. *Endocrinology.* 2002;143:4084–4095.
78. Strauss BL, Pittman R, Pixley MR, Nilson JH, Boime I. Expression of the beta subunit of chorionic gonadotropin in transgenic mice. *J Biol Chem.* 1994;269:4968–4973.
79. Risma KA, Hirshfield AN, Nilson JH. Elevated luteinizing hormone in prepubertal transgenic mice causes hyperandrogenemia, precocious puberty, and substantial ovarian pathology. *Endocrinology.* 1997;138:3540–3547.
80. Keri RA, Lozada KL, Abdul-Karim FW, Nadeau JH, Nilson JH. Luteinizing hormone induction of ovarian tumors: oligogenic differences between mouse strains dictates tumor disposition. *Proc Natl Acad Sci USA.* 2000;97:383–387.
81. Couse JF, Yates MM, Sanford R, Nyska A, Nilson JH, Korach KS. Formation of cystic ovarian follicles associated with elevated luteinizing hormone requires estrogen receptor-beta. *Endocrinology.* 2004;145:4693–4702.
82. Flaws JA, Abbud R, Mann RJ, Nilson JH, Hirshfield AN. Chronically elevated luteinizing hormone depletes primordial follicles in the mouse ovary. *Biol Reprod.* 1997;57:1233–1237.
83. Mann RJ, Keri RA, Nilson JH. Transgenic mice with chronically elevated luteinizing hormone are infertile due to anovulation, defects in uterine receptivity, and midgestation pregnancy failure. *Endocrinology.* 1999;140:2592–2601.
84. Mann RJ, Keri RA, Nilson JH. Consequences of elevated luteinizing hormone on diverse physiological systems: use of the LHbetaCTP transgenic mouse as a model of ovarian hyperstimulation-induced pathophysiology. *Recent Prog Horm Res.* 2003;58: 343–375.

85. Kero J, Poutanen M, Zhang FP, Rahman N, McNicol AM, Nilson JH, Keri RA, Huhtaniemi IT. Elevated luteinizing hormone induces expression of its receptor and promotes steroidogenesis in the adrenal cortex. *J Clin Invest.* 2000;105:633–641.
86. Mohammad HP, Abbud RA, Parlow AF, Lewin JS, Nilson JH. Targeted overexpression of luteinizing hormone causes ovary-dependent functional adenomas restricted to cells of the Pit-1 lineage. *Endocrinology.* 2003;144:4626–4636.
87. Strauss BL, Boime I. Cellular localization of the human chorionic gonadotropin beta-subunit in transgenic mouse placenta. *Endocrinology.* 2000;141:430–437.
88. Huhtaniemi I, Rulli S, Ahtiainen P, Poutanen M. Multiple sites of tumorigenesis in transgenic mice overproducing hCG. *Mol Cell Endocrinol.* 2005;234:117–126.
89. Huhtaniemi I, Ahtiainen P, Pakarainen T, Rulli SB, Zhang FP, Poutanen M. Genetically modified mouse models in studies of luteinising hormone action. *Mol Cell Endocrinol.* 2006;252:126–135.
90. Ahtiainen P, Sharp V, Rulli SB, Rivero-Muller A, Mamaeva V, Roytta M, Huhtaniemi I. Enhanced LH action in transgenic female mice expressing hCGbeta-subunit induces pituitary prolactinomas; the role of high progesterone levels. *Endocr Relat Cancer.* 2010;17:611–621.
91. Ratner LD, Stevens G, Bonaventura MM, Lux-Lantos VA, Poutanen M, Calandra RS, Huhtaniemi IT, Rulli SB. Hyperprolactinemia induced by hCG leads to metabolic disturbances in female mice. *J Endocrinol.* 2016;230:157–169.
92. Ratner LD, Gonzalez B, Ahtiainen P, Di Giorgio NP, Poutanen M, Calandra RS, Huhtaniemi IT, Rulli SB. Short-term pharmacological suppression of the hyperprolactinemia of infertile hCG-overproducing female mice persistently restores their fertility. *Endocrinology.* 2012;153:5980–5992.
93. Kuorelahti A, Rulli S, Huhtaniemi I, Poutanen M. Human chorionic gonadotropin (hCG) up-regulates wnt5b and wnt7b in the mammary gland, and hCGbeta transgenic female mice present with mammary gland tumors exhibiting characteristics of the Wnt/beta-catenin pathway activation. *Endocrinology.* 2007;148:3694–3703.
94. Gonzalez B, Ratner LD, Scerbo MJ, Di Giorgio NP, Poutanen M, Huhtaniemi IT, Calandra RS, Lux-Lantos VA, Cambiasso MJ, Rulli SB. Elevated hypothalamic aromatization at the onset of precocious puberty in transgenic female mice hypersecreting human chorionic gonadotropin: effect of androgens. *Mol Cell Endocrinol.* 2014;390:102–111.
95. Yarram SJ, Perry MJ, Christopher TJ, Westby K, Brown NL, Lamminen T, Rulli SB, Zhang FP, Huhtaniemi I, Sandy JR, Mansell JP. Luteinizing hormone receptor knockout (LuRKO) mice and transgenic human chorionic gonadotropin (hCG)-overexpressing mice (hCG alphabeta+) have bone phenotypes. *Endocrinology.* 2003;144:3555–3564.
96. Rulli SB, Ahtiainen P, Makela S, Toppari J, Poutanen M, Huhtaniemi I. Elevated steroidogenesis, defective reproductive organs, and infertility in transgenic male mice overexpressing human chorionic gonadotropin. *Endocrinology.* 2003;144:4980–4990.
97. Ahtiainen P, Rulli SB, Shariatmadari R, Pelliniemi LJ, Toppari J, Poutanen M, Huhtaniemi IT. Fetal but not adult Leydig cells are susceptible to adenoma formation in response to persistently high hCG level: a study on hCG overexpressing transgenic mice. *Oncogene.* 2005;24:7301–7309.
98. Shenker A. Activating mutations of the lutropin choriogonadotropin receptor in precocious puberty. *Receptors Channels.* 2002;8:3–18.
99. Shenker A, Laue L, Kosugi S, Merendino Jr JJ, Minegishi T, Cutler Jr GB. A constitutively activating mutation of the luteinizing hormone receptor in familial male precocious puberty. *Nature.* 1993;365:652–654.
100. Wu SM, Leschek EW, Rennert OM, Chan WY. Luteinizing hormone receptor mutations in disorders of sexual development and cancer. *Front Biosci.* 2000;5:D343–D352.

101. Mason AJ, Hayflick JS, Zoeller RT, Young 3rd WS, Phillips HS, Nikolics K, Seeburg PH. A deletion truncating the gonadotropin-releasing hormone gene is responsible for hypogonadism in the hpg mouse. *Science.* 1986;234:1366–1371.
102. Mason AJ, Pitts SL, Nikolics K, Szonyi E, Wilcox JN, Seeburg PH, Stewart TA. The hypogonadal mouse: reproductive functions restored by gene therapy. *Science.* 1986;234:1372–1378.
103. Ma X, Dong Y, Matzuk MM, Kumar TR. Targeted disruption of luteinizing hormone beta-subunit leads to hypogonadism, defects in gonadal steroidogenesis, and infertility. *Proc Natl Acad Sci USA.* 2004;101:17294–17299.
104. Oury F, Ferron M, Huizhen W, Confavreux C, Xu L, Lacombe J, Srinivas P, Chamouni A, Lugani F, Lejeune H, Kumar TR, Plotton I, et al. Osteocalcin regulates murine and human fertility through a pancreas-bone-testis axis. *J Clin Invest.* 2013;123:2421–2433.
105. Bernard DJ, Tran S. Mechanisms of activin-stimulated FSH synthesis: the story of a pig and a FOX. *Biol Reprod.* 2013;88:78.
106. Coss D, Mellon PL, Thackray VG. A FoxL in the Smad house: activin regulation of FSH. *Trends Endocrinol Metab.* 2010;21:562–568.
107. Thackray VG, Mellon PL, Coss D. Hormones in synergy: regulation of the pituitary gonadotropin genes. *Mol Cell Endocrinol.* 2010;314:192–203.
108. Kumar TR, Fairchild-Huntress V, Low MJ. Gonadotrope-specific expression of the human follicle-stimulating hormone beta-subunit gene in pituitaries of transgenic mice. *Mol Endocrinol.* 1992;6:81–90.
109. Kumar TR, Low MJ. Gonadal steroid hormone regulation of human and mouse follicle stimulating hormone beta-subunit gene expression in vivo. *Mol Endocrinol.* 1993;7:898–906.
110. Kumar TR, Low MJ. Hormonal regulation of human follicle-stimulating hormone-beta subunit gene expression: GnRH stimulation and GnRH-independent androgen inhibition. *Neuroendocrinology.* 1995;61:628–637.
111. Kumar TR, Schuff KG, Nusser KD, Low MJ. Gonadotroph-specific expression of the human follicle stimulating hormone beta gene in transgenic mice. *Mol Cell Endocrinol.* 2006;247:103–115.
112. Kumar TR, Palapattu G, Wang P, Woodruff TK, Boime I, Byrne MC, Matzuk MM. Transgenic models to study gonadotropin function: the role of follicle-stimulating hormone in gonadal growth and tumorigenesis. *Mol Endocrinol.* 1999;13:851–865.
113. Allan CM, Handelsman DJ. Transgenic models for exploring gonadotropin biology in the male. *Endocrine.* 2005;26:235–239.
114. Allan CM, Haywood M, Swaraj S, Spaliviero J, Koch A, Jimenez M, Poutanen M, Levallet J, Huhtaniemi I, Illingworth P, Handelsman DJ. A novel transgenic model to characterize the specific effects of follicle-stimulating hormone on gonadal physiology in the absence of luteinizing hormone actions. *Endocrinology.* 2001;142:2213–2220.
115. Haywood M, Spaliviero J, Jimemez M, King NJ, Handelsman DJ, Allan CM. Sertoli and germ cell development in hypogonadal (hpg) mice expressing transgenic follicle-stimulating hormone alone or in combination with testosterone. *Endocrinology.* 2003;144: 509–517.
116. Allan CM, Wang Y, Jimenez M, Marshan B, Spaliviero J, Illingworth P, Handelsman DJ. Follicle-stimulating hormone increases primordial follicle reserve in mature female hypogonadal mice. *J Endocrinol.* 2006;188:549–557.
117. McTavish KJ, Jimenez M, Walters KA, Spaliviero J, Groome NP, Themmen AP, Visser JA, Handelsman DJ, Allan CM. Rising follicle-stimulating hormone levels with age accelerate female reproductive failure. *Endocrinology.* 2007;148:4432–4439.
118. Allan CM, Kalak R, Dunstan CR, McTavish KJ, Zhou H, Handelsman DJ, Seibel MJ. Follicle-stimulating hormone increases bone mass in female mice. *Proc Natl Acad Sci USA.* 2010;107:22629–22634.

119. Miller WL, Shafiee-Kermani F, Strahl BD, Huang HJ. The nature of FSH induction by GnRH. *Trends Endocrinol Metab.* 2002;13:257–263.
120. Huang HJ, Sebastian J, Strahl BD, Wu JC, Miller WL. Transcriptional regulation of the ovine follicle-stimulating hormone-beta gene by activin and gonadotropin-releasing hormone (GnRH): involvement of two proximal activator protein-1 sites for GnRH stimulation. *Endocrinology.* 2001;142:2267–2274.
121. Huang HJ, Sebastian J, Strahl BD, Wu JC, Miller WL. The promoter for the ovine follicle-stimulating hormone-beta gene (FSHbeta) confers FSHbeta-like expression on luciferase in transgenic mice: regulatory studies in vivo and in vitro. *Endocrinology.* 2001;142:2260–2266.
122. Pernasetti F, Spady TJ, Hall SB, Rosenberg SB, Givens ML, Anderson S, Paulus M, Miller WL, Mellon PL. Pituitary tumorigenesis targeted by the ovine follicle-stimulating hormone beta-subunit gene regulatory region in transgenic mice. *Mol Cell Endocrinol.* 2003;203:169–183.
123. Markkula M, Kananen K, Paukku T, Mannisto A, Loune E, Frojdman K, Pelliniemi LJ, Huhtaniemi I. Induced ablation of gonadotropins in transgenic mice expressing herpes simplex virus thymidine kinase under the FSH beta-subunit promoter. *Mol Cell Endocrinol.* 1995;108:1–9.
124. Kumar TR, Wang Y, Lu N, Matzuk MM. Follicle stimulating hormone is required for ovarian follicle maturation but not male fertility. *Nat Genet.* 1997;15:201–204.
125. Wreford NG, Kumar TR, Matzuk MM, de Kretser DM. Analysis of the testicular phenotype of the follicle-stimulating hormone beta-subunit knockout and the activin type II receptor knockout mice by stereological analysis. *Endocrinology.* 2001;142: 2916–2920.
126. Tapanainen JS, Aittomaki K, Min J, Vaskivuo T, Huhtaniemi IT. Men homozygous for an inactivating mutation of the follicle-stimulating hormone (FSH) receptor gene present variable suppression of spermatogenesis and fertility. *Nat Genet.* 1997;15:205–206.
127. Aittomaki K, Herva R, Stenman UH, Juntunen K, Ylostalo P, Hovatta O, de la Chapelle A. Clinical features of primary ovarian failure caused by a point mutation in the follicle-stimulating hormone receptor gene. *J Clin Endocrinol Metab.* 1996;81:3722–3726.
128. Aittomaki K, Lucena JL, Pakarinen P, Sistonen P, Tapanainen J, Gromoll J, Kaskikari R, Sankila EM, Lehvaslaiho H, Engel AR, et al. Mutation in the follicle-stimulating hormone receptor gene causes hereditary hypergonadotropic ovarian failure. *Cell.* 1995;82:959–968.
129. Combelles CM, Carabatsos MJ, Kumar TR, Matzuk MM, Albertini DF. Hormonal control of somatic cell oocyte interactions during ovarian follicle development. *Mol Reprod Dev.* 2004;69:347–355.
130. El-Hayek S, Clarke HJ. Follicle-stimulating hormone increases gap junctional communication between somatic and germ-line follicular compartments during murine oogenesis. *Biol Reprod.* 2015;93:47.
131. El-Hayek S, Demeestere I, Clarke HJ. Follicle-stimulating hormone regulates expression and activity of epidermal growth factor receptor in the murine ovarian follicle. *Proc Natl Acad Sci USA.* 2014;111:16778–16783.
132. Abel MH, Huhtaniemi I, Pakarinen P, Kumar TR, Charlton HM. Age-related uterine and ovarian hypertrophy in FSH receptor knockout and FSHbeta subunit knockout mice. *Reproduction.* 2003;125:165–173.
133. Iqbal J, Sun L, Kumar TR, Blair HC, Zaidi M. Follicle-stimulating hormone stimulates TNF production from immune cells to enhance osteoblast and osteoclast formation. *Proc Natl Acad Sci USA.* 2006;103:14925–14930.
134. Sun L, Peng Y, Sharrow AC, Iqbal J, Zhang Z, Papachristou DJ, Zaidi S, Zhu LL, Yaroslavskiy BB, Zhou H, Zallone A, Sairam MR, et al. FSH directly regulates bone mass. *Cell.* 2006;125:247–260.

135. Zaidi M, Blair HC, Iqbal J, Zhu LL, Kumar TR, Zallone A, Sun L. Proresorptive actions of FSH and bone loss. *Ann NY Acad Sci.* 2007;1116:376–382.
136. Stilley JA, Christensen DE, Dahlem KB, Guan R, Santillan DA, England SK, Al-Hendy A, Kirby PA, Segaloff DL. FSH receptor (FSHR) expression in human extragonadal reproductive tissues and the developing placenta, and the impact of its deletion on pregnancy in mice. *Biol Reprod.* 2014;91:74.
137. Cui H, Zhao G, Liu R, Zheng M, Chen J, Wen J. FSH stimulates lipid biosynthesis in chicken adipose tissue by upregulating the expression of its receptor FSHR. *J Lipid Res.* 2012;53:909–917.
138. Radu A, Pichon C, Camparo P, Antoine M, Allory Y, Couvelard A, Fromont G, Hai MT, Ghinea N. Expression of follicle-stimulating hormone receptor in tumor blood vessels. *N Engl J Med.* 2010;363:1621–1630.
139. Siraj A, Desestret V, Antoine M, Fromont G, Huerre M, Sanson M, Camparo P, Pichon C, Planeix F, Gonin J, Radu A, Ghinea N. Expression of follicle-stimulating hormone receptor by the vascular endothelium in tumor metastases. *BMC Cancer.* 2013;13:246.
140. Kumar TR, Low MJ, Matzuk MM. Genetic rescue of follicle-stimulating hormone beta-deficient mice. *Endocrinology.* 1998;139:3289–3295.
141. Wang H, Larson M, Jablonka-Shariff A, Pearl CA, Miller WL, Conn PM, Boime I, Kumar TR. Redirecting intracellular trafficking and the secretion pattern of FSH dramatically enhances ovarian function in mice. *Proc Natl Acad Sci USA.* 2014;111: 5735–5740.
142. Arey BJ, Lopez FJ. Are circulating gonadotropin isoforms naturally occurring biased agonists? Basic and therapeutic implications. *Rev Endocr Metab Disord.* 2011;12:275–288.
143. Fares F. The role of O-linked and N-linked oligosaccharides on the structure-function of glycoprotein hormones: development of agonists and antagonists. *Biochim Biophys Acta.* 2006;1760:560–567.
144. Lambert A, Talbot JA, Anobile CJ, Robertson WR. Gonadotrophin heterogeneity and biopotency: implications for assisted reproduction. *Mol Hum Reprod.* 1998;4:619–629.
145. Smitz J, Wolfenson C, Chappel S, Ruman J. Follicle-stimulating hormone: a review of form and function in the treatment of infertility. *Reprod Sci.* 2016;23:706–716.
146. Ulloa-Aguirre A, Timossi C, Damian-Matsumura P, Dias JA. Role of glycosylation in function of follicle-stimulating hormone. *Endocrine.* 1999;11:205–215.
147. Bousfield GR, Butnev VY, Butnev VY, Hiromasa Y, Harvey DJ, May JV. Hypo-glycosylated human follicle-stimulating hormone (hFSH(21/18)) is much more active in vitro than fully-glycosylated hFSH (hFSH(24)). *Mol Cell Endocrinol.* 2014;382: 989–997.
148. Bousfield GR, Butnev VY, Rueda-Santos MA, Brown A, Hall AS, Harvey DJ. Macro- and micro-heterogeneity in pituitary and urinary follicle-stimulating hormone glycosylation. *J Glycomics Lipidomics.* 2014;4.
149. Bousfield GR, Butnev VY, Walton WJ, Nguyen VT, Huneidi J, Singh V, Kolli VS, Harvey DJ, Rance NE. All-or-none N-glycosylation in primate follicle-stimulating hormone beta-subunits. *Mol Cell Endocrinol.* 2007;260–262:40–48.
150. Bousfield GR, Butnev VY, White WK, Hall AS, Harvey DJ. Comparison of Follicle-stimulating hormone glycosylation microheterogenity by quantitative negative mode nano-electrospray mass spectrometry of peptide-N glycanase-released oligosaccharides. *J Glycomics Lipidomics.* 2015;5(1):5.
151. Butnev VY, Butnev VY, May JV, Shuai B, Tran P, White WK, Brown A, Smalter Hall A, Harvey DJ, Bousfield GR. Production, purification, and characterization of recombinant hFSH glycoforms for functional studies. *Mol Cell Endocrinol.* 2015;405:42–51.
152. Jiang C, Hou X, Wang C, May JV, Butnev VY, Bousfield GR, Davis JS. Hypoglycosylated hFSH has greater bioactivity than fully glycosylated recombinant hFSH in human granulosa cells. *J Clin Endocrinol Metab.* 2015;100:E852–E860.

153. Walton WJ, Nguyen VT, Butnev VY, Singh V, Moore WT, Bousfield GR. Characterization of human FSH isoforms reveals a nonglycosylated beta-subunit in addition to the conventional glycosylated beta-subunit. *J Clin Endocrinol Metab.* 2001;86:3675–3685.
154. Wang H, Butnev V, Bousfield GR, Kumar TR. A human FSHB transgene encoding the double N-glycosylation mutant (asn(7delta) asn(24delta)) FSHbeta subunit fails to rescue Fshb null mice. *Mol Cell Endocrinol.* 2016;426:113–124.
155. Macias MJ, Martin-Malpartida P, Massague J. Structural determinants of Smad function in TGF-beta signaling. *Trends Biochem Sci.* 2015;40:296–308.
156. Massague J. How cells read TGF-beta signals. *Nat Rev Mol Cell Biol.* 2000;1:169–178.
157. Massague J. TGFbeta signalling in context. *Nat Rev Mol Cell Biol.* 2012;13:616–630.
158. Massague J, Gomis RR. The logic of TGFbeta signaling. *FEBS Lett.* 2006;580:2811–2820.
159. Massague J, Seoane J, Wotton D. Smad transcription factors. *Genes Dev.* 2005;19: 2783–2810.
160. Shi Y, Massague J. Mechanisms of TGF-beta signaling from cell membrane to the nucleus. *Cell.* 2003;113:685–700.
161. Chang H, Brown CW, Matzuk MM. Genetic analysis of the mammalian transforming growth factor-beta superfamily. *Endocr Rev.* 2002;23:787–823.
162. Chang H, Lau AL, Matzuk MM. Studying TGF-beta superfamily signaling by knock-outs and knockins. *Mol Cell Endocrinol.* 2001;180:39–46.
163. Matzuk MM, Kumar TR, Vassalli A, Bickenbach JR, Roop DR, Jaenisch R, Bradley A. Functional analysis of activins during mammalian development. *Nature.* 1995;374: 354–356.
164. Matzuk MM, Kumar TR, Shou W, Coerver KA, Lau AL, Behringer RR, Finegold MJ. Transgenic models to study the roles of inhibins and activins in reproduction, oncogenesis, and development. *Recent Prog Horm Res.* 1996;51:123–154.
165. Ethier JF, Findlay JK. Roles of activin and its signal transduction mechanisms in reproductive tissues. *Reproduction.* 2001;121:667–675.
166. Gaddy-Kurten D, Tsuchida K, Vale W. Activins and the receptor serine kinase super-family. *Recent Prog Horm Res.* 1995;50:109–129.
167. Woodruff TK. Regulation of cellular and system function by activin. *Biochem Pharmacol.* 1998;55:953–963.
168. Lau AL, Kumar TR, Nishimori K, Bonadio J, Matzuk MM. Activin betaC and betaE genes are not essential for mouse liver growth, differentiation, and regeneration. *Mol Cell Biol.* 2000;20:6127–6137.
169. Lau AL, Nishimori K, Matzuk MM. Structural analysis of the mouse activin beta C gene. *Biochim Biophys Acta.* 1996;1307:145–148.
170. Fortin J, Ongaro L, Li Y, Tran S, Lamba P, Wang Y, Zhou X, Bernard DJ. Minireview: activin signaling in gonadotropes: what does the FOX say to the SMAD? *Mol Endocrinol.* 2015;29:963–977.
171. Pangas SA, Woodruff TK. Activin signal transduction pathways. *Trends Endocrinol Metab.* 2000;11:309–314.
172. Matzuk MM, Kumar TR, Bradley A. Different phenotypes for mice deficient in either activins or activin receptor type II. *Nature.* 1995;374:356–360.
173. Lotinun S, Pearsall RS, Horne WC, Baron R. Activin receptor signaling: a potential therapeutic target for osteoporosis. *Curr Mol Pharmacol.* 2012;5:195–204.
174. Matzuk MM, Finegold MJ, Su JG, Hsueh AJ, Bradley A. Alpha-inhibin is a tumour-suppressor gene with gonadal specificity in mice. *Nature.* 1992;360:313–319.
175. Matzuk MM, Finegold MJ, Mather JP, Krummen L, Lu H, Bradley A. Development of cancer cachexia-like syndrome and adrenal tumors in inhibin-deficient mice. *Proc Natl Acad Sci USA.* 1994;91:8817–8821.

176. Kumar TR, Wang Y, Matzuk MM. Gonadotropins are essential modifier factors for gonadal tumor development in inhibin-deficient mice. *Endocrinology.* 1996;137: 4210–4216.
177. Nagaraja AK, Agno JE, Kumar TR, Matzuk MM. Luteinizing hormone promotes gonadal tumorigenesis in inhibin-deficient mice. *Mol Cell Endocrinol.* 2008;294:19–28.
178. Bernard DJ, Burns KH, Haupt B, Matzuk MM, Woodruff TK. Normal reproductive function in InhBP/p120-deficient mice. *Mol Cell Biol.* 2003;23:4882–4891.
179. Bernard DJ, Chapman SC, Woodruff TK. Inhibin binding protein (InhBP/p120), betaglycan, and the continuing search for the inhibin receptor. *Mol Endocrinol.* 2002;16:207–212.
180. Chapman SC, Bernard DJ, Jelen J, Woodruff TK. Properties of inhibin binding to betaglycan, InhBP/p120 and the activin type II receptors. *Mol Cell Endocrinol.* 2002;196:79–93.
181. Ma X, Reyna A, Mani SK, Matzuk MM, Kumar TR. Impaired male sexual behavior in activin receptor type II knockout mice. *Biol Reprod.* 2005;73:1182–1190.
182. Vassalli A, Matzuk MM, Gardner HA, Lee KF, Jaenisch R. Activin/inhibin beta B subunit gene disruption leads to defects in eyelid development and female reproduction. *Genes Dev.* 1994;8:414–427.
183. Kumar TR, Varani S, Wreford NG, Telfer NM, de Kretser DM, Matzuk MM. Male reproductive phenotypes in double mutant mice lacking both FSHbeta and activin receptor IIA. *Endocrinology.* 2001;142:3512–3518.
184. Fortin J, Boehm U, Weinstein MB, Graff JM, Bernard DJ. Follicle-stimulating hormone synthesis and fertility are intact in mice lacking SMAD3 DNA binding activity and SMAD2 in gonadotrope cells. *FASEB J.* 2014;28:1474–1485.
185. Fortin J, Boehm U, Deng CX, Treier M, Bernard DJ. Follicle-stimulating hormone synthesis and fertility depend on SMAD4 and FOXL2. *FASEB J.* 2014;28:3396–3410.
186. Sandoval-Guzman T, Gongrich C, Moliner A, Guo T, Wu H, Broberger C, Ibanez CF. Neuroendocrine control of female reproductive function by the activin receptor ALK7. *FASEB J.* 2012;26:4966–4976.
187. Tsuchida K, Nakatani M, Yamakawa N, Hashimoto O, Hasegawa Y, Sugino H. Activin isoforms signal through type I receptor serine/threonine kinase ALK7. *Mol Cell Endocrinol.* 2004;220:59–65.
188. Zhou X, Wang Y, Ongaro L, Boehm U, Kaartinen V, Mishina Y, Bernard DJ. Normal gonadotropin production and fertility in gonadotrope-specific Bmpr1a knockout mice. *J Endocrinol.* 2016;229:331–341.
189. Dong J, Albertini DF, Nishimori K, Kumar TR, Lu N, Matzuk MM. Growth differentiation factor-9 is required during early ovarian folliculogenesis. *Nature.* 1996;383: 531–535.
190. Choi SG, Wang Q, Jia J, Pincas H, Turgeon JL, Sealfon SC. Growth differentiation factor 9 (GDF9) forms an incoherent feed-forward loop modulating follicle-stimulating hormone beta-subunit (FSHbeta) gene expression. *J Biol Chem.* 2014;289:16164–16175.
191. Bilezikjian LM, Blount AL, Corrigan AZ, Leal A, Chen Y, Vale WW. Actions of activins, inhibins and follistatins: implications in anterior pituitary function. *Clin Exp Pharmacol Physiol.* 2001;28:244–248.
192. Guo Q, Kumar TR, Woodruff T, Hadsell LA, DeMayo FJ, Matzuk MM. Overexpression of mouse follistatin causes reproductive defects in transgenic mice. *Mol Endocrinol.* 1998;12:96–106.
193. Matzuk MM, Lu N, Vogel H, Sellheyer K, Roop DR, Bradley A. Multiple defects and perinatal death in mice deficient in follistatin. *Nature.* 1995;374:360–363.
194. Lin SY, Morrison JR, Matzuk MM, de Kretser DM. Spermatogenesis does not require the local production of follistatin. *Reproduction.* 2006;132:601–605.

195. Jorgez CJ, Klysik M, Jamin SP, Behringer RR, Matzuk MM. Granulosa cell-specific inactivation of follistatin causes female fertility defects. *Mol Endocrinol.* 2004;18: 953–967.
196. Meehan TP, Narayan P. Constitutively active luteinizing hormone receptors: consequences of in vivo expression. *Mol Cell Endocrinol.* 2007;260–262:294–300.
197. Narayan P. Genetic models for the study of luteinizing hormone receptor function. *Front Endocrinol.* 2015;6:152.
198. Allan CM, Lim P, Robson M, Spaliviero J, Handelsman DJ. Transgenic mutant D567G but not wild-type human FSH receptor overexpression provides FSH-independent and promiscuous glycoprotein hormone Sertoli cell signaling. *Am J Physiol Endocrinol Metab.* 2009;296:E1022–E1028.
199. Peltoketo H, Strauss L, Karjalainen R, Zhang M, Stamp GW, Segaloff DL, Poutanen M, Huhtaniemi IT. Female mice expressing constitutively active mutants of FSH receptor present with a phenotype of premature follicle depletion and estrogen excess. *Endocrinology.* 2010;151:1872–1883.
200. Lei ZM, Mishra S, Zou W, Xu B, Foltz M, Li X, Rao CV. Targeted disruption of luteinizing hormone/human chorionic gonadotropin receptor gene. *Mol Endocrinol.* 2001;15:184–200.
201. Zhang FP, Poutanen M, Wilbertz J, Huhtaniemi I. Normal prenatal but arrested postnatal sexual development of luteinizing hormone receptor knockout (LuRKO) mice. *Mol Endocrinol.* 2001;15:172–183.
202. Abel MH, Wootton AN, Wilkins V, Huhtaniemi I, Knight PG, Charlton HM. The effect of a null mutation in the follicle-stimulating hormone receptor gene on mouse reproduction. *Endocrinology.* 2000;141:1795–1803.
203. Dierich A, Sairam MR, Monaco L, Fimia GM, Gansmuller A, LeMeur M, Sassone-Corsi P. Impairing follicle-stimulating hormone (FSH) signaling in vivo: targeted disruption of the FSH receptor leads to aberrant gametogenesis and hormonal imbalance. *Proc Natl Acad Sci USA.* 1998;95:13612–13617.
204. Danilovich N, Sairam MR. Targeting gonadotropin receptor genes: reproductive biology, aging, and related health implications. *Endocrine.* 2005;26:219–226.
205. Hasuwa H, Ueda J, Ikawa M, Okabe M. miR-200b and miR-429 function in mouse ovulation and are essential for female fertility. *Science.* 2013;341:71–73.
206. Topilko P, Schneider-Maunoury S, Levi G, Trembleau A, Gourdji D, Driancourt MA, Rao CV, Charnay P. Multiple pituitary and ovarian defects in Krox-24 (NGFI-A, Egr-1)-targeted mice. *Mol Endocrinol.* 1998;12:107–122.
207. Tran S, Zhou X, Lafleur C, Calderon MJ, Ellsworth BS, Kimmins S, Boehm U, Treier M, Boerboom D, Bernard DJ. Impaired fertility and FSH synthesis in gonadotrope-specific Foxl2 knockout mice. *Mol Endocrinol.* 2013;27:407–421.
208. Zhao L, Bakke M, Krimkevich Y, Cushman LJ, Parlow AF, Camper SA, Parker KL. Steroidogenic factor 1 (SF1) is essential for pituitary gonadotrope function. *Development.* 2001;128:147–154.
209. Zhao L, Bakke M, Parker KL. Pituitary-specific knockout of steroidogenic factor 1. *Mol Cell Endocrinol.* 2001;185:27–32.
210. Bliss SP, Miller A, Navratil AM, Xie J, McDonough SP, Fisher PJ, Landreth GE, Roberson MS. ERK signaling in the pituitary is required for female but not male fertility. *Mol Endocrinol.* 2009;23:1092–1101.
211. Wang H, Graham I, Hastings R, Gunewardena S, Brinkmeier ML, Conn PM, Camper SA, Kumar TR. Gonadotrope-specific deletion of Dicer results in severely suppressed gonadotropins and fertility defects. *J Biol Chem.* 2015;290:2699–2714.
212. Cacioppo JA, Koo Y, Lin PC, Osmulski SA, Ko CD, Ko C. Generation of an estrogen receptor beta-iCre knock-in mouse. *Genesis.* 2016;54:38–52.

213. Gieske MC, Kim HJ, Legan SJ, Koo Y, Krust A, Chambon P, Ko C. Pituitary gonadotroph estrogen receptor-alpha is necessary for fertility in females. *Endocrinology.* 2008;149:20–27.
214. Sanchez-Criado JE, Trudgen K, Millan Y, Blanco A, Monterde J, Garrido-Gracia JC, Gordon A, Aguilar R, de Las Mulas JM, Ko C. Estrogen receptor (ESR) 2 partially offsets the absence of ESR1 in gonadotropes of pituitary-specific Esr1 knockout female mice. *Reproduction.* 2012;143:549–558.
215. Wu S, Chen Y, Fajobi T, DiVall SA, Chang C, Yeh S, Wolfe A. Conditional knockout of the androgen receptor in gonadotropes reveals crucial roles for androgen in gonadotropin synthesis and surge in female mice. *Mol Endocrinol.* 2014;28:1670–1681.
216. Kanchiswamy CN, Maffei M, Malnoy M, Velasco R, Kim JS. Fine-tuning next-generation genome editing tools. *Trends Biotechnol.* 2016;34:562–574.
217. Lin J, Musunuru K. Genome engineering tools for building cellular models of disease. *FEBS J.* 2016. [Epub ahead of print].
218. Miles LA, Garippa RJ, Poirier JT. Design, execution, and analysis of pooled in vitro CRISPR/Cas9 screens. *FEBS J.* 2016. [Epub ahead of print].
219. Wang H, La Russa M, Qi LS. CRISPR/Cas9 in genome editing and beyond. *Annu Rev Biochem.* 2016;85:227–264.
220. Xiong X, Chen M, Lim WA, Zhao D, Qi LS. Genome engineering and disease research. *Annu Rev Genomics Hum Genet.* 2016. [Epub ahead of print].
221. Davis JS, Kumar TR, May JV, Bousfield GR. Naturally occurring follicle-stimulating hormone glycosylation variants. *J Glycomics Lipidomics.* 2014;4:e117.
222. de Waal E, Yamazaki Y, Ingale P, Bartolomei MS, Yanagimachi R, McCarrey JR. Gonadotropin stimulation contributes to an increased incidence of epimutations in ICSI-derived mice. *Hum Mol Genet.* 2012;21:4460–4472.
223. Laan M, Grigorova M, Huhtaniemi IT. Pharmacogenetics of follicle-stimulating hormone action. *Curr Opin Endocrinol Diabetes Obes.* 2012;19:220–227.
224. Nataraja SG, Yu HN, Palmer SS. Discovery and development of small molecule allosteric modulators of glycoprotein hormone receptors. *Front Endocrinol.* 2015;6:142.
225. Oduwole OO, Huhtaniemi IT. Feasibility of male hormonal contraception: lessons from clinical trials and animal experiments. *Curr Mol Pharmacol.* 2014;7:109–118.
226. Zdrojewicz Z, Konieczny R, Papier P, Szten F. Brdt bromodomains inhibitors and other modern means of male contraception. *Adv Clin Exp Med.* 2015;24:705–714.

CHAPTER THREE

Clinical Applications of Gonadotropins in the Female: Assisted Reproduction and Beyond

L. Casarini*[†], G. Brigante*[†,‡,1], M. Simoni*[†,‡], D. Santi*[†,‡]

[*] Unit of Endocrinology, Department of Biomedical, Metabolic and Neural Sciences, University of Modena and Reggio Emilia, Modena, Italy
[†] Center for Genomic Research, University of Modena and Reggio Emilia, Modena, Italy
[‡] Azienda USL of Modena, Modena, Italy
[1] Corresponding author. E-mail address: giulia.brigante@unimore.it

Contents

Progress in Molecular Biology and Translational Science, Volume 143
ISSN 1877-1173
http://dx.doi.org/10.1016/bs.pmbts.2016.08.002

Abstract

Gonadotropins (LH, FSH, and hCG) act in concert in the regulation of female reproductive system. Exploiting this influence, they are part of the assisted reproductive technique protocols. In this review we analyze the effectiveness of the different available gonadotropin formulations and the consequent adverse events. Moreover, different protocols for poor-responders and polycystic ovary syndrome affected women are explored. All these clinical different approaches have specific molecular bases, covered in this review starting from evolution and population genetics, getting to in vitro studies of gonadotropins action. Beyond their application in assisted reproductive technique, gonadotropins have also been largely studied for their intertwined network of interactions with other hormones, which all together contribute to the functioning of the reproductive system and other hormonal axes. In particular, there is both clinical and molecular evidence of interaction between thyroid hormones and insulin growth factors with gonadotropins. Finally, gonadotropins are widely studied for their role in the maintenance of the proper balance between cell proliferation and differentiation, and therefore in cancer.

1. INTRODUCTION

The concerted action of gonadotropins plays a pivotal role in both male and female reproductive system.[1,2] Although featured by similar molecular structures, follicle-stimulating hormone (FSH), luteinizing hormone (LH), and chorionic gonadotropin (hCG) specifically bind to their receptors and mediate specific intracellular signaling cascades in the target cells. FSH is a heterodimeric glycoprotein secreted by the anterior pituitary gland, together with LH, under the control of hypothalamic gonadotropin-releasing hormone (GnRH). These hormones share the α-chain with other glycoprotein hormones, such as the thyroid-stimulating hormone (TSH), encoded by the unique *CGA* gene. On the contrary, the β-subunits, even if structurally similar, are encoded by specific genes and differ among hormones conferring action specificity.

Gonadotropins are available as therapeutic drugs, used in several clinical conditions and in particular in the setting of assisted reproductive technology (ART).[3] First ART cycle dates back to the 1970s, when in vitro fertilization (IVF) was developed to achieve pregnancy in case of female and/or male subfertility/infertility. Currently, a gold standard approach to

ART does not exist. Several ART schemes and different, not properly evidence-based, approaches to oocyte fertilization and embryo implantation are available. In this complex setting, several gonadotropin preparations are commercially available, derived from either human urine or manufactured by recombinant technology.[4] However, the clinical use of gonadotropins is reported to be variably associated to different outcomes in ART. This review aims to evaluate differences in gonadotropin preparations and ART schemes, with particular attention to possible adverse events and hormones interaction.

2. MOLECULAR BIOLOGY OF THE OVARIAN RESPONSE

The ovarian response is finely coordinated by gonadotropins, which vary in serum concentration levels and isoform composition during menstrual cycle and pregnancy. These variations result in potentially different ligand-receptor interactions and physiological modulations of growth-promoting and differentiation signals in their target cells, exerted in tandem with steroid hormones and paracrine factors. These mechanisms are overall evolutionarily conserved among mammals, although signals of endocrine adaptation and modification exist across the species. Especially, the primate-specific, dual ligand system for the same receptor provided by LH and hCG yields a refined control of folliculogenesis and pregnancy, respectively.[5] Moreover, single-nucleotide polymorphisms (SNPs) within gonadotropin and their receptor genes modulate gonadal physiology. This results in different models of ovarian cycle,[6,7] spermatogenesis,[8,9] and pregnancy.[10,11]

2.1 Concepts From Evolution and Population Genetics in Females

In mammals, FSH and LH are released by the pituitary under GnRH stimulus, although species-specific differences characterize the ovarian cycle duration and estrus seasonality. The recruitment of one to several dominant follicles occurs by the action of FSH and estrogens, before ovulation under LH induction. The glycoprotein hormone receptors are G protein-coupled receptors including the TSH receptor (TSHR), the LH/CG receptor (LHCGR), and the FSH receptor (FSHR). They possess an extracellular, ligand-specific domain,[12] connected to the 7 transmembrane stretches by a hinge region, and an intracellular portion interacting with molecular mediators of signal transduction[13] (Fig. 1). Several SNPs fall within the FSH β-subunit *(FSHB)* and *FSHR* genes, resulting in protein variants modulating the reproductive physiological mechanisms in humans.[14] SNPs consist in single nucleotide changes falling within the gene sequences. Especially, an SNP at the C-terminal portion of FSHR affects the ovarian response to FSH and endocrine parameters. This is the nucleotide substitution "A" to "G" at position 2039 (c.2039A>G) of the *FSHR* gene, resulting in the residue change Asparagine to Serine at position 680 of the receptor amino acid chain (p.N680S; rs6166, in the National Center for

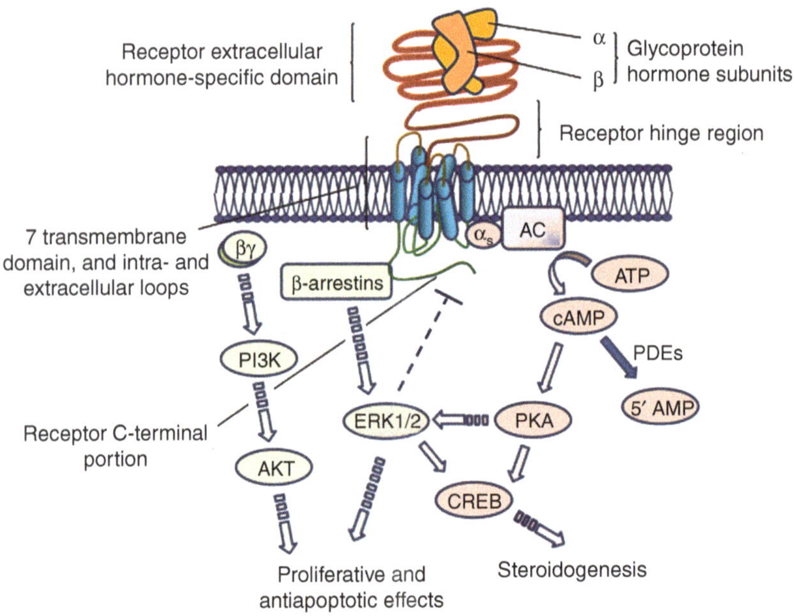

Figure 1 Main intracellular signaling pathways mediated by glycoprotein hormone receptors. Hormone binding to the extracellular portion induces the conformational change of the receptor and recruitment of several intracellular interactors. The activation of the canonical cAMP/PKA-pathway depends on ATP conversion to cAMP by adenylyl cyclase *(AC)* enzyme, which in turn, is activated by Gα_s protein. Intracellular cAMP increase is controlled by the phosphodiesterase enzymes *(PDEs)* and results in downstream ERK1/2 and CREB phosphorylation, thus mediating proliferative events and receptor downregulation, and steroidogenesis, respectively. Later ERK1/2 recruitment may occur via alternative interactors, that is, β-arrestins, which mediates receptor internalization. The βγ subunits of G proteins modulate the phosphatidylinositol-4, 5-bisphosphate 3-kinase (PIP3K)/AKT-pathway, simultaneously to the aforementioned signaling cascades, resulting in antiapoptotic events.

Biotechnology Information SNP database; http://www.ncbi.nlm.nih.gov). This polymorphism is in strong linkage disequilibrium, at least in Caucasians, with another SNP (p.A307T) in the extracellular domain of the receptor, resulting in two main discrete receptor isoforms FSHR 307A/680N and 307T/680S. The Serine FSHR variant (p.N680S S variant) retains lower sensitivity to FSH than the Asparagine variant (p.N680S N variant). Indeed p.N680S homozygous S women have higher mean serum FSH levels and require superior doses of exogenous hormone for assisted reproduction than p.N680S N carriers.[15] In spite of the lower FSH sensitivity, the FSHR p.N680S polymorphism is common among humans, displaying an allele frequency ranging from less than 30% among Eastern Asians and Oceanians to more than 50% in populations of Central Asian ancestry.[16] The high prevalence of both alleles suggests that different ovarian sensitivity to FSH, due to the receptor variant, is not decisively linked to subfertility in humans, rather resulting in different features of the ovarian cycle, such as the duration. The peculiar combinations of SNPs within the *FSHR* and *FSHB* genes in haplotypes might affect some reproductive parameters,[17] such as low, but prolonged, ovulatory frequency and higher age at menopause, presumably not affecting the individual reproductive success and propagating these genotypes. However, the relevance of the ovarian cycle features, for example, the menstrual cycle length, in determining human reproductive success is still unclear. Maybe natural selection pressure has been more determinant in selecting phenotypes characterized by advantageous pregnancy features and high live birth success.

Pregnancy is not sustained by hCG in nonprimate species, where placental tissue invasion and embryo growth are regulated by other hormones. The β-subunit genes of chorionic gonadotropin exist only in primates (except equids) and are encoded by *CGB* gene copies likely originated by sequential duplications of an ancient *LHB* gene, and belonging to a relatively unstable genomic region.[18] Interestingly, climbing the phylogenic tree of primates toward humans an increasing number of *CGB* gene copies occur, resulting in 1–3 genes among the early primates, 3–5 in chimpanzees, gorillas, and orangutans, and 8 in Homo sapiens. The number of *CGB* genes expressed and the amount of N- and O-linked oligosaccharide structures associated with chorionic gonadotropin isoforms, resulting in a wide variety of acidic variants in humans, reflect probably the high level of complexity of placentation and invasiveness described. Moreover, hCG regulates several events as progesterone production, trophoblast invasion, angiogenesis, and local immune tolerance. Taken together, the dual LH/CG system provides an example of endocrine adaptation to higher energy demand during pregnancy, which is likely required for a high brain mass development in humans.[11] This implies that, in humans, LH and hCG are two naturally occurring ligands for the same receptor, which is able to discriminate between them and modulates specific, different physiological effects. The extracellular domain of LHCGR retains different, LH- or hCG-specific binding sites. Moreover, the hinge region is fundamental to mediate ligand-specific intracellular signaling. It is encoded by the *LHCGR* exon 10, out of 11 in total, and is implied in the transmission of the conformational change, induced by the hormone binding at the extracellular domain, to the transmembrane region. The importance of exon 10 in discriminating between the LH- and hCG-induced signaling was revealed by a case report describing a hypogonadal male patient not responsive to endogenous LH.[19] The genomic deletion of exon 10 was found in this man affected by Leydig cell hypoplasia type II, delayed pubertal development and small testicles. In this patient, testosterone production was induced by the administration of exogenous hCG, revealing differential, ligand-specific interaction and cell signaling mediated by LHCGR. Moreover, the new world monkey *Callithrix jacchus* is naturally expressing an *LHCGR* lacking exon 10, and the production of a peculiar pituitary chorionic gonadotropin that partially replaces LH activity in this species.[20] These findings point out the importance of the amino acids falling at the hinge region of the receptor for correct function of the dual LH/CG system. Most importantly, the variety of receptor isoforms and the wide number of ligand types and variants reveal the complexity of the adaptation process to reproduction occurred during primate evolution.

2.2 Modulation of FSH Action by FSHR Variants In Vitro

Although the effects of the p.N680S *FSHR* SNP in the modulation of ovarian response were found more than a decade ago,[21] the evidence that the receptor variants are linked to different response to FSH in vitro has been lacking for a long time. A study evaluating the binding affinity and the production of the second messenger cAMP mediated by the two *FSHR* variants in transiently transfected COS-7 cells found no significant results.[22] In these experiments, cells were incubated with increasing FSH doses and cAMP was measured after few hours, not allowing the evaluation of the kinetics of the second messenger accumulation over time. These results were later independently confirmed in transfected 293T cells, detecting no different phosphatidyl inositol turnover mediated by the two receptor isoforms.[23] Similarly, another study evaluating FSHR trans-activation by high TSH doses found no different cAMP production in transfected HEK293 cells.[24] More recently, a detailed comparison between the cAMP production, progesterone, and estradiol synthesis mediated by the two receptor isoforms in vitro were performed using human primary granulosa-lutein cells.[25] Homozygous *FSHR* p.N680S N or S cells were obtained from genotyped patients undergoing oocyte retrieval for assisted reproduction. Granulosa cells spontaneously express gonadotropin receptors, thus providing a more physiological model than transfected cell lines. However, since no significantly different results were found again, the results obtained in vitro did not support the several clinical observations. We reasoned that this discrepancy may be due to the achievement of a steady metabolic plateau of the cells.[26] Indeed, the studies described earlier were performed under saturation conditions, by treating cells for relatively long times, when each endpoint was evaluated after reaching similar, maximal levels of activation disregarding the different kinetics. In fact, when different *FSHR* p.N680S N- and S-mediated responses were evaluated in short-term (15 min) experiments[27] in granulosa cells, FSH dose-responses revealed different patterns of extracellular-regulated kinases (ERK1/2) and cAMP response elements binding protein (CREB) phosphorylation, depending on the receptor isoform. Especially, the p.N680S S *FSHR* mediates a lower cAMP/protein kinase A (PKA)-dependent pERK1/2 and pCREB activation than the N isoform. The different ERK1/2 phosphorylation is linked to a slower, FSHR p.N680S S-mediated accumulation of cAMP over 2 h, despite achieving the same plateau levels of the other receptor isoform, resulting in a weaker downstream but more prolonged CREB phosphorylation and progesterone synthesis in vitro. These results elucidate the molecular mechanism by which the two common FSHR variants have different sensitivity to FSH and, reasonably, may modulate ovarian response. We could speculate that the different kinetics of steroid production observed in in vitro experiments may be reflected in a diverse inhibitory feedback mechanism on FSH production at the pituitary level in vivo. Moreover, this knowledge may be useful to improve the therapeutic approaches for ART by developing new molecules acting as modulators of the kinetics of FSHR-mediated signaling.

2.3 Differentiating LH and hCG Action In Vitro

Several molecular and biochemical features differentiate LH and hCG. Aside from their aforementioned β-subunit encoding genes and differently related oligosaccharide structures, the two hormones differ in their half-life (about 60 min for LH, several hours for hCG).[28] The long clearance of hCG is due to the extension of 24 amino acids of its carboxyl-terminal peptide (CTP). Moreover, hCG seems to have a higher binding affinity than LH for their common receptor,[29] although more detailed experiments evaluating the displacement of radio-labeled ligands should be provided to confirm this issue. On the other hand, the two hormones bind distinct sites located in the extracellular domain of the human receptor, which, in turn, is able to differentiate the ligand-specific signaling through interaction with sulfated tyrosine residues of the hinge region and some extracellular loops.[30] These studies indicate a possible mechanism by which LHCGR differently mediates intracellular signaling cascades depending on the ligand, a property called biased signaling and known for many G protein-coupled receptors.[31]

Follicular maturation is a physiological process relying on proliferative events regulated by gonadotropins, estrogens, and several paracrine factors. Reflecting the role of LH during folliculogenesis, the treatment of cultured human granulosa cells in vitro by picomolar doses of this hormone resulted in the activation of proliferative and antiapoptotic, rather than steroidogenic intracellular signals, which are amplified by FSH cotreatment.[32,33] Especially, the action of LH is exerted through the ERK1/2- and protein kinase B (AKT)-pathways, which are well-known, essential key regulators of the follicular functions[34,35] (Fig. 1), resulting in the rescue of the cells from apoptosis in vitro. The ERK-mediated proliferative activity of LH was demonstrated also in caprine granulosa cells, where the prolonged exposure to the hormone promotes proliferation, while hCG induces cAMP-dependent decreasing of cell growth.[36] The treatment of human granulosa cells with hCG in vitro induces a slower, but more potent cAMP production than LH, resulting in higher downstream progesterone synthesis and decreased cell viability. We could speculate that the high steroidogenic potential of hCG serves to fulfill the high demand of progesterone during pregnancy, but the translation of in vitro data to human physiology requires to be further investigated. cAMP-mediated steroidogenesis is strictly connected with proapoptotic events mediated by hCG in vitro, likely as a consequence of the tumor protein 53 (p53) activation by high intracellular levels of the second messenger.[37] Taken together, in vitro data indicate differential proliferative and antiapoptotic activity of LH and highly steroidogenic effect of hCG, which may reflect their different physiological roles in vivo.

3. GONADOTROPINS IN ASSISTED REPRODUCTION

3.1 Urinary Gonadotropin Preparations

Urinary products include human menopausal gonadotropin (hMG), urinary FSH (u-FSH), and urinary hCG (u-hCG). HMG was the first gonadotropin preparation used in the treatment of infertility since the 1960s and, successively, in controlled ovarian stimulation (COS) during ART. This compound was firstly purified from urine of postmenopausal or pregnant women, and it contained both FSH and LH activities.[1] LH activity in hMG is variable, deriving from both LH and hCG molecules within the same preparation.[1,38] In particular, hMG compounds provide LH activity through residual LH molecules and supplementation of u-hCG in the final products.[1,38] The LH-related activity residual of the first hMG compounds has been the reason of the high batch-to-batch inconsistency of these preparations. Therefore, the purification techniques progressively improved, reducing the differences between batches in new hMG formulations.

Alongside hMG production, various urinary-derived gonadotropin specimens were developed. In 1983 the first u-FSH (urofollitropin) was developed, in which only less than 5% of total protein showed FSH activity.[39] According to the further development of the purification process, a highly purified (HP) u-FSH is now available, with over 80–95% of FSH content.[40] Differently to hMG compounds, HP-u-FSH shows a higher batch-to-batch consistency.[41] Obtaining reproducible u-FSH batches represents an important challenge in the gonadotropin purification process. In order to obtain reproducible u-FSH batches, a complex purification process has been developed. It provides several consecutive purification steps, starting from hormone absorption and elution, to anion/cation exchange, and hydrophobic chromatography.[42] Currently, the purification process remains different among gonadotropin preparations. Anyway, hCG and several impurities could remain in the final purified products, leading to differences among batches.[42] In particular, leucocyte elastase inhibitor, protein C inhibitor, zinc-glycoprotein, growth factors, glycoproteins, binding proteins, insulin-like growth factor binding protein-7, transferrin, and immunoglobulins could be present in the final compound.[42–44] The advent of new technologies, such as reverse-phase high-performance liquid chromatography analysis and 2D gel electrophoresis, reduces the concentration of these protein impurities in urinary compounds.[1,42] This phase of purification is extremely important because the biological activity of contaminants may potentially expose patients to adverse events.[43] Recent proteomic analyses detected the presence in hMG compounds of prions,[45,46] which are brain protein isoforms associated to transmissible spongiform encephalopathy.[47] In spite of these sporadic reports, there is no evidence whatsoever of adverse effects in women treated with hMG for ART.[48] Thus, considering the pharmacovigilance evidence and the clinical outcome, current urinary gonadotropin compounds used in ART are safe.

A, perhaps, more challenging issue in urinary preparations is the method of quantification of the gonadotropin-related activity. This quantification is extremely important in clinical practice and it is generally simple for u-FSH and u-hCG, since these compounds could be calibrated against standard FSH and hCG activity, respectively. Activity quantification is more challenging for hMG, considering the different extractive origin, the different glycoprotein contents, and the longer half-life of hCG compared to LH.[49] Since hMG contains both LH and hCG molecules, with a usually higher amount of hCG, the increased hMG purity is associated to a lower LH content.[42] In fact, the addition of hCG molecules from external sources allows the maintainance of the final required FSH:LH ratio of 1:1.[4,42] Moreover, in vivo bioassays for LH activity detection were originally calibrated against standard hCG, and not against standard LH.[49] Thus, the quantification of LH activity in mixed preparations, such as hMG is challenging.

3.2 Recombinant Gonadotropin Preparations

Recombinant preparations are available for FSH (r-FSH), LH (r-LH), and hCG (r-hCG). In particular, follitropin α, follitropin β, lutropin α, choriogonadotropin α, and chorifollitorpin α were developed using Chinese hamster ovary (CHO) cell lines.[50] In 1988 the first r-FSH was produced, characterized by high availability, bioactivity, and purity, without any LH or other urinary contaminants.[51,52] The recombinant process used for manufacturing FSH was further improved, using either filled-by-bioassay or filled-by-mass technologies. These new techniques allow obtaining r-FSH with a higher batch-to-batch consistency compared to earlier urinary-derived preparations.[4] Moreover, recombinant technologies allow to quantify gonadotropin activity in molar terms.[53]

The increased purity and consistency of recombinant compounds allows the study of the role of gonadotropin isoforms, especially regarding FSH. In fact, human pituitary FSH consists of three or four major isoforms, differing in the glycosylation of the β-chain.[54] In vitro studies demonstrated a greater bioactivity of hypoglycosylated FSH, compared to fully glycosylated isoforms.[55] Thus, the age-dependent decrease in hypoglycosylated FSH could in part explain the reduction in the ovarian responsiveness.[55] Considering the commercially available preparations, u- and r-FSH differ because the latter shows minor differences in the structure of the carbohydrate side chains and contains more basic and fewer acidic isohormones.[56–58] However it is not possible to categorize these structural differences in specific glycosylation pattern. Indeed, both u- and r-FSH result in poor forms of hypoglycosylated isoforms.[54,55] In order to ensure the glycosylation pattern of endogenous gonadotropins, a recent r-FSH compound has been produced using human cell line of fetal retinal origin.[59]

Considering the two available formulations of r-FSH, several studies compared their efficacy in ART. The total FSH bioactivity delivered from a single dose preparation after reconstitution differs between follitropin α and follitropin β.[4] However, no differences were found between α and β follitropins, in terms of number of follicles, oocytes retrieved, and fertilized oocytes.[60] In addition, corifollitropin α is the first hybrid FSH produced, demonstrating a prolonged FSH-activity.[61] The β subunit of this molecule contains the hCG CTP, giving a longer circulation half-life and extending time to peak levels.[61] Clinical trials showed a similar ovarian response and safety profile in ART cycles between corifollitropin α and follitropin α.[61]

On the other hand, lutropin α and choriogonadotropin α have been extensively characterized using traditional physicochemical, biological, and immunological techniques (e.g., terminal amino acid sequencing, peptide mapping, electrophoresis, chromatographic analysis, bioassay, and so on) as well as state-of-the-art analytical procedures (e.g., mass spectrometry).[28] As well as for FSH, both lutropin α and choriogonadotropin α are a mixture of isoforms and present N- and C-terminal heterogeneity. The overall characterization of these molecules has been extensively done, concluding that they are structurally and biologically comparable to endogenous gonadotropins, although not strictly identical, due to slight differences in post-translational modifications.[28]

3.3 Biosimilar Gonadotropins

A biosimilar drug is a product that is similar but not identical to the DNA-derived, originator biological product.[62] Since the technological details used for the originator product, for example, culture technology, media, cell lines, and purification systems, etc., are not publicly available and considering that it is not possible to apply exactly the same procedure to manufacture the product, some difference in the final biological product should be assumed. In the case of gonadotropins, this might result in changes in the microheterogeneity of the biological compound, with possible different effects in the clinical results.[39]

Considering that biotechnological products derive from a complex, multistep process, biosimilars must be characterized by a thorough production process control.[39] As for the originator, each step could directly affect purity, quality, or biological activity of the final DNA-derived product.[39] The European Medicine Agency (EMA) establishes guidelines for biosimilar development, requiring a full quality dossier for each compound, demonstrating the physicochemical and biological similarity and clinical noninferiority to the reference originator.[39]

Recombinant FSH biosimilars must have not only the same amino-acid sequence of the originator, but also may have only minor differences in posttranslational forms or product-related impurities.[39] This quality control is necessary to reduce impurities, which could have adverse consequences in the clinical practice.[39] Each biosimilar product should demonstrate similar efficacy and safety, without immunogenicity, in clinical studies designed to detect possible differences in response between the biosimilar and the originator.[39]

3.4 Do Different Preparations Have Different Clinical Effects?

The choice of the best gonadotropin to be used in ART is a very controversial field of medicine, since no real, nonsponsored, evidence-based comparative trial has ever been performed and independently confirmed. On the other hand, given the psychological pressure of infertile couples to achieve pregnancy in the shortest possible time, often in a battle against ovarian ageing due to the steadily increasing age of the woman age in couples undergoing ART, performing truly randomized clinical trials in this field is almost impossible. In general, EMA guidelines recommend demonstrating the clinical efficacy of a urinary or recombinant gonadotropin in a randomized, double blind, prospective clinical trial, designed with appropriate statistical power in the setting of ART. Moreover, EMA guidelines suggest to consider the number of retrieved oocytes as primary outcome. However, several reports demonstrated that pregnancy rate adjusted for oocytes retrieved is a better outcome to compare the efficacy and the equivalence between different gonadotropin preparations.[63]

Considering the different origin of gonadotropin preparations, one of the most important issues is the comparison between urinary and recombinant preparations. In particular, the use of chromatographic techniques to assess the potency of urinary compounds shows inconstant variation in their composition.[52] U-FSH, for example, shows higher acidity, heterogeneity, glycosylation, and sialylation levels than r-FSH.[64] This discrepancy is explained by the differences in the enzymatic pool between CHO and human pituitary cells, leading to a variable degree in glycosylation pattern of the final proteins. Although in vitro models demonstrated a higher bioactivity of less acidic isoforms, no clear demonstration of glycosylation influence on FSH action in vivo is currently known. The comparison between u-FSH and r-FSH does not show any difference considering various outcomes, from oocytes retrieved to pregnancy rate.[65] However, this comparison needs to be carefully evaluated, considering that couple baseline conditions are predictive of ART outcome and these studies are extremely heterogeneous in terms of inclusion/exclusion criteria, drug used, and regimen chosen. Informative results were provided by metaanalysis, which stated no differences between urinary and recombinant compounds.[1,38,65] Recently, the Cochrane Collaboration confirmed the similar efficacy of r-FSH and u-FSH, in terms of number of oocytes retrieved and live birth rate.[66]

On the other hand, hCG is considered ideal for follicular development because of its long half-life. On the contrary, r-LH has a short half-life and requires daily injections of higher doses to sustain follicle development.[1] However, LH and hCG showed different effects during COS,[1] especially when considered together with FSH. This difference in clinical practice is still not clear, considering the heterogeneity of studies available in the literature.[1]

4. OVARIAN STIMULATION PROTOCOLS IN ART

The COS protocol is the most delicate and variable phase in ART, aimed at increasing the number of follicles and mature oocytes. A wide literature on various pharmacological approaches to COS is available, literally "inventing" new ways and alternative schemes in the empirical attempt to improve IVF outcome. Both urinary and recombinant FSH can be used alone or associated with LH activity, which, in turn, can be provided in the form of hMG, r-LH, or hCG.[67] Moreover, the use of steroid hormones, such as progesterone, estrogen, testosterone or dehydroepiandrosterone, or other hormones, such as growth hormone, was considered in several researches.[67] However, the latter compounds do not show a significant improvement of pregnancy rate in clinical trials and their use remains under debate.

4.1 Classic Scheme(s) of Gonadotropin Stimulation in ART

A standardized scheme in ART does not exist and empirically personalized therapy remains the most used approach. Currently, demographic and anthropomorphic data are the primary source of information to guide the clinician's choice. This information is completed by functional, hormonal, and genetic biomarkers. Among these, antimüllerian hormone (AMH) represents a functional marker of ovarian reserve.[68] On the other hand, antral follicular count (AFC), representing the number of follicles maturing in each cycle, provides a prediction of COS yield.[68] In this context, the patient genetic profile defines the underlying physiology, completing the predictive role of functional and hormonal biomarkers.[68] However, despite the wide number of predictive tools available, there is currently no consensus in the literature about the best COS scheme to be used. In fact, the multidimensional variability of women undergoing ART renders a single treatment approach unsuitable, leaving the personalization of the treatment the only effective approach to improve pregnancy rate and ART outcomes.

Fig. 2 summarizes the ovarian stimulation phase during ART. During the classic COS scheme, the first step is represented by the prevention of the occurrence of spontaneous LH surge, which may happen before the leading follicle reaches the optimum diameter for triggering ovulation. This down-regulation phase is carried out using a GnRH agonist or antagonist. The GnRH agonist is administered before starting gonadotropins ovarian stimulation. Physiologically, GnRH is released in a cyclical fashion to stimulate the gonadotropin release, whereas this continuous, exogenous administration used in COS has the opposite effect, causing a cessation of physiological gonadotropin production.[69] If a GnRH antagonist is chosen, this is administered in the mid-follicular phase, after the start of gonadotropin stimulation and prior to triggering final maturation of oocytes.[70]

The second phase of COS provides the gonadotropins administration, using FSH alone or combined to LH activity. The use of FSH for COS is

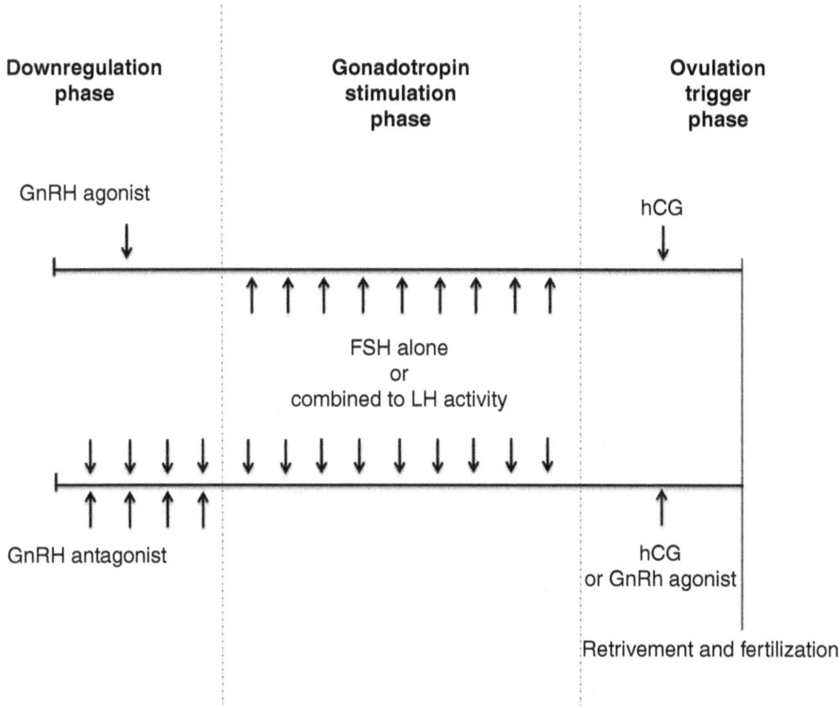

Figure 2 Different schemes of controlled ovarian stimulation. Standard controlled ovarian stimulation cycles are described. The downregulation phase is represented on the left, the gonadotropin stimulation in the middle, and the ovulation trigger phase on the right, dividing according to GnRH agonist or antagonist protocol. *FSH*, Follicle-stimulating hormone; *GnRH*, gonadotropin-releasing hormone; *hCG*, human chorionic gonadotropin; *LH*, luteinizing hormone.

necessary for multiple follicular growth, although deciding which type of FSH preparation and dose remains arbitrary.[71–73] In particular, despite some suggestions of an increased cumulative pregnancy rate using r-FSH, data available in the literature suggest a substantial interchangeability between urinary and recombinant FSH preparations, in terms of efficacy and safety.[74,75] Similarly, the use of daily short-acting FSH administration or long-acting FSH formulation is equally effective.[76]

In the classic COS scheme, the FSH dose generally ranges from 150 IU daily, for high responders, 225–300 IU daily, for normal responders, up to 375 IU, for poor responders.[67] However, the treatment should be evaluated for each patient and the FSH dose adapted to prognostic factors and demographic characteristics. Moreover, the gonadotropin dose is usually adjusted during the COS phase, according to hormonal and ultrasonography parameters. Recently, the CONSORT study proposed an algorithm to individualize r-FSH doses in ART.[77] This prospective, multicenter pilot study enrolling normoresponder women, aged between 18 and 34 years, proposed a mathematical FSH dose increase of 37.5 IU daily, according to basal FSH serum levels, body mass index, age, and AFC.[77] Currently, the CONSORT calculator represents an example of tools that could be used to recommend the starting FSH dose[78,79] albeit physicians are reluctant to use schemes that were obtained in standardized women, since they are not corresponding to "real life." Therefore, the so-called "personalized" approach to ovarian stimulation currently used in clinical practice is left, in the majority of cases, to the clinical experience and personal feelings of the treating doctor rather than based on solid evidence.

During COS, LH activity is often added to increase FSH efficacy. However, several doubts remain regarding the need to add LH activity to FSH and also about the best LH activity-containing preparation to be used in COS. In fact, differences between LH and hCG are well demonstrated in in vitro models,[32] while direct comparison of LH and hCG in human clinical trials has been very limited so far.[1] Moreover, the use of hMG together with FSH poses several challenges, since the actual amount of LH and hCG molecules in hMG preparations varies considerably. Thus, there is an ongoing controversy in ART whether LH or hCG should be used for COS together with FSH or not.

Finally, the last phase is the oocyte triggering (Fig. 2). This phase aims to stimulate the oocytes grown during COS to go through the last stage of maturation, ready to be retrieved and fertilized in vitro. This final oocyte maturation is usually achieved by high dose hCG administration, although several studies suggested a negative hCG effect on endometrial receptivity and embryo quality and the danger of ovarian hyperstimulation syndrome (OHSS) occurrence.[80,81] Currently, GnRH agonists, stimulating an endogenous LH peak, represent an alternative to hCG trigger.[82] However, this approach is applicable only in COS schemes using GnRH antagonists in the downregulation phase.[82] Although this protocol can be used in selected women, the GnRH agonist trigger is associated to lower OHSS incidence, without differences in live birth rate.[83]

4.2 ART in Poor Responder Women

Poor responders are those women in whom the classic COS results in few oocytes retrieved, reduced number of embryos for transfer, and, consequently, unsatisfactory pregnancy rates. One of the most important factors influencing COS response is the woman's age, confirmed to be the best predictor of female fertility also in ART. Increasing age inevitably results in substantial germ cell loss, particularly important by the age of 40,[84] although it does not affect all women equally.

The real prevalence of poor responders is difficult to estimate due to a variety of definitions. The ESHRE workshop held in Bologna (Italy) in 2010 reached the consensus in the definition of criteria for this subgroup of women. However, despite this new definition and classification, the individualization of gonadotropin type and dose remains the key point even in this subgroup of women. In particular, poor responders and women over 35 years might benefit from exogenous LH administration during COS, together with FSH.[85] Lehert et al. recently showed a higher number of oocytes retrieved in poor responders treated with r-LH and r-FSH, compared to those treated with r-FSH alone.[86] Although the individualization of gonadotropin treatment remains the most used approach in ART, the combined treatment with FSH and LH or hCG in poor responder women might improve the final ART outcome, but this remains to be convincingly demonstrated.

4.3 ART in Women With Polycystic Ovarian Syndrome

Stimulation protocols for women affected by polycystic ovary syndrome (PCOS) present some peculiar aspects. PCOS is the most common endocrine disorder affecting 5–15% of women of fertile-age worldwide. PCOS is characterized by variable phenotypes, featured by hyperandrogenism, ovulatory dysfunction, and polycystic ovarian morphology, occurring in different combinations, and associated with different grades of metabolic disease risk.[87] Moreover, insulin resistance is linked to PCOS and affects ovulation by disrupting the hypothalamic-pituitary-ovarian axis. Due to the polygenic nature of the disorder, and its several environmental modulators (e.g., lifestyle, dietary habits, and so on), the management of PCOS women in assisted reproduction has to be personalized. Clomiphene citrate, an antagonist/modulator of the estrogen receptor, may support ovulation in PCOS women, and is recommended as a primary infertility treatment in these patients. Moreover, treatment with metformin decreases the risk of complications, such as metabolic syndrome and gestational diabetes.[88,89] Finally, there are no specific guidelines on the use of gonadotropins for infertility in PCOS women, and additional drugs are variably used in a personalized regimen, depending on the patient phenotype and clinician's attitude.

In conclusion, the personalization of COS schemes represents a well-established clinical practice and is believed to offer several benefits,[90] although this assumption has never been substantiated by really solid and well-controlled clinical studies. Future research should aim to find the best therapeutic strategies, considering all available schemes, tuning the gonadotropin type and dose on the basis of potential ovarian response in each woman[90] in well-controlled settings.

5. ADVERSE EVENTS IN OVARIAN STIMULATION

5.1 Ovarian Hyperstimulation Syndrome

OHSS is an iatrogenic complication of COS. It results from the complications of an exaggerated ovarian response to gonadotropin administration. The incidence of moderate to severe OHSS is between 3.1% and 8% of in vitro fertilization (IVF) cycles but can be as high as 20% in high-risk women.[91] Several studies have been conducted to understand what are the factors that point out a woman as more at risk than another, and some pretreatment factors, such as low body mass index, young age, high estradiol or interleukins levels, renin-angiotensin system activation, and vascular endothelial growth factor expression have been associated to OHSS.[92] In addition, history of PCOS or previous OHSS raise the risk, as well as high pretreatment basal AMH concentrations and high AFC at ultrasound examination.[91] Among factors occurring during COS, rapidly rising estrogen levels, a large number of developing follicles on the day of hCG administration, and a large number of oocytes retrieved are considered risk factors for developing OHSS.[91] However, none of the cited factors have been shown to be independently predictive of OHSS and it must be noted that women without any risk factors can develop OHSS as well.

Some genetic markers have been considered. In particular, the association between OHSS and alterations in gonadotropin receptor genes have been investigated, but no consistent correlations were found yet. In fact, some rare activating mutations of *FSHR* gene result in a predisposition to OHSS.[93] However, mutations in the *FSHR* are rare in OHSS and SNPs could rather be involved. So far, 744 SNPs have been identified in the *FSHR* gene and only 8 are located in the coding region. In particular, the P.N680S variant is a predictor of the severity of symptoms in patients who develop OHSS.[94] Even if the number of patients investigated so far is too limited to draw any conclusion, the ovarian response to gonadotropins seems to be dependent on *FSHR* genotype and the identification of the genetic profile prior to ovarian stimulation could permit a safer treatment.

There is currently no consensus on the pathophysiology of OHSS. However, hCG is considered the most critical mediator of the syndrome. As a confirmation of this, OHSS does not develop when hCG is not used during COS and, moreover, the increased hCG exposure is associated with an increased risk of OHSS.[91] In fact, OHSS has two distinct clinical presentations: an "early" and a "late" manifestation[91] (Fig. 3). The "early" OHSS occurs within 9 days of hCG administration as ovulatory trigger and is the effect of exogenous hCG on ovaries already stimulated by gonadotropins. "Late" OHSS occurs more than 10 days after the use of hCG and represents the ovarian response to endogenous hCG produced by the trophoblast.

The role of hCG in the pathophysiology of OHSS is attributable to the mediation of the release of vascular endothelial growth factor-A, that promotes angiogenesis and vascular hyperpermeability.[95] Another pathophysiological mechanism mediated by hCG is the activation of the intraovarian

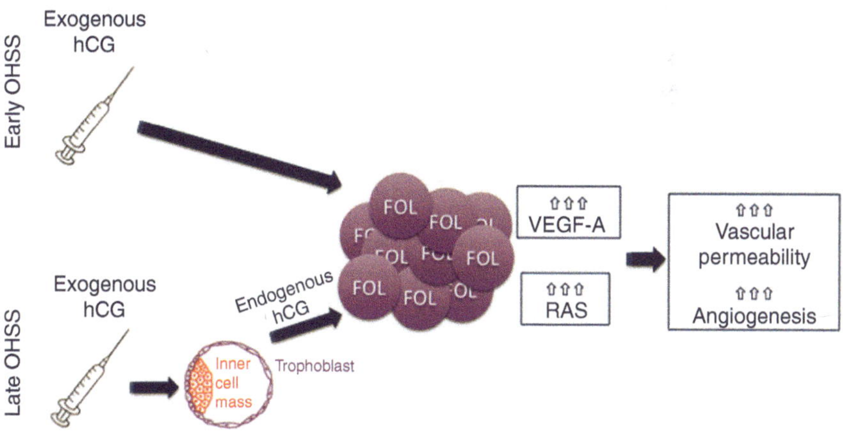

Figure 3 Early and late clinical presentations of ovarian hyperstimulation syndrome *(OHSS)*. "Early" and "late" clinical presentations of OHSS. The "early" manifestation occurs within 9 days of human chorionic gonadotropin *(hCG)* administration and is the effect of exogenous hCG on ovarian follicles *(FOL)*. The "late" manifestation of OHSS occurs more than 10 days after the use of hCG and represents the ovarian response to endogenous hCG produced by the trophoblast. They both result in the release of vascular endothelial growth factor-A *(VEGF-A)* and in the activation of the intraovarian renin angiotensin system *(RAS)* that promotes angiogenesis and vascular hyperpermeability.

renin angiotensin system, which regulates vascular permeability, angiogenesis, endothelial proliferation, and prostaglandin release.[96] Accordingly, OHSS clinical manifestation is characterized by extravasation of fluid from the intravascular compartment into the third space. And the resulting haemoconcentration leads to complications, such as hypercoagulability and reduced end-organ perfusion.[97]

In women who are considered at a high risk of OHSS, treatment regimens must be modified in order to avoid an excessive ovarian response. Interventions with evidence of effect in reducing OHSS include the use of metformin in women with PCOS, the use of GnRH antagonists rather than agonists, and the use of GnRH agonist triggers in GnRH antagonist stimulation cycles.[98] Moreover, reducing the dose of FSH, freezing all embryos, and transferring a single embryo are considered appropriate interventions to reduce OHSS.

In conclusion, OHSS is a serious condition that requires further research to develop optimal prevention strategies and facilitate the early detection of women at risk.

5.2 Premature Ovarian Failure

Ovarian reserve indicates the woman's reproductive potential, in terms of follicular and oocyte number and quality. It is well established that ovarian reserve generally declines with age. This decline begins already in utero but becomes particularly relevant when the total remaining number of follicles in the ovaries is near to 25,000 and there is an accelerated loss of follicles together with a qualitative change in the remaining follicles.[99] These remaining follicles show fewer granulosa cells, diminished production of steroids and glycoproteins, decreased mitosis and increased apoptosis.[100] The final result is an alteration in the endocrine, paracrine, and autocrine pathways, leading to an altered communication between granulosa cells and oocytes.[101] The rate of ovarian reserve decline varies among individuals and depends on medical history, environmental and genetic factors. However, several diseases could affect ovarian reserve, such as endometriosis, pelvic inflammatory disease, systemic illness, chemotherapy, and smoking. In general, the incidence of poor ovarian response ranges from 10 to 20% and the prevalence varies depending on the woman's age, with a lower prevalence among women younger than 34 years and significantly increasing with age, reaching 50% in women aged 43–44 years.[102] Poor response could be suspected in the presence of high FSH and estradiol levels and low AFC and AMH.[90] AMH cut-off levels for poor ovarian response are reported between 0.1 and 2.97 ng/mL, considering different cut-off ranges among assays.

Several studies proposed that poor performance in ART might be considered the first detectable sign of impending menopause.[103] Nikolaou et al. performed a retrospective cohort study demonstrating a strong association between an extremely poor response to COS and premature ovarian failure.[104] In this way, the poor response to gonadotropin stimulation should be considered an early sign of ovarian failure.

5.3 Other Complications

In general, poor responders or women with PCOS may develop adverse obstetric outcomes after COS, such as ectopic pregnancy, cervical incompetence, preeclampsia, birth of children small for gestational age.[90] However, the most frequent ART complications are cycle cancellation, prolonged treatment, increased treatment burden and low pregnancy rates.[90] On the other hand, it is well known that COS does not increase the gynaecological cancer risk, as well as hormone-dependent and hormone-independent endometrial and breast cancer.[105]

6. GONADOTROPINS BEYOND ART

6.1 Gonadotropins and Thyroid Axis

Thyroid hormones (TH) play an important role during prereproductive, reproductive, and postreproductive age both in males and females. They are produced by the thyroid under the stimulus of TSH. Thyroxine (T4) is the main TH synthesized by the thyroid and can be locally converted to the more biologically active triiodothyronine (T3) by three types of deiodinase (DIO) enzymes. At the tissue level, T3 binds to nuclear receptors, TR-α and TR-β, regulating the transcription of target genes. In addition, TH can also have a rapid effect, mediated by the binding to integrin $\alpha\nu\beta3$ receptor, activating the MAPK/ERK1/2 signal-transduction cascade.[106]

TH exert their earlier influence at the embryogenetic level, regulating sexual differentiation and gonadal development in mammalian and nonmammalian species.[107] In particular, TH modulate the expression of sex-determining genes, the production of gonadotropins in the hypothalamus-pituitary-gonad axis, and the production of sexual steroids through direct and indirect regulation of steroidogenic enzyme expression and activity.[107]

Thereafter, TH influence hypothalamic secretion of GnRH, which is the main regulator of puberty onset.[108] In parallel, thyroid volume and function increase as an adaptation to body and sexual development.[109] Thus, a cross-talk between thyroid and the gonadotropin axis is particularly relevant during periods of hormonal turmoil, like puberty. It is known that hypothyroidism occurring before puberty can cause a delay in sexual maturity,[109] possibly because of prolactin (PRL) increase. In fact, hypothyroidism induces an elevation of the hypothalamic thyrotropin-releasing hormone (TRH), known to stimulate both PRL and TSH. As a consequence, the secondary hyperprolactinemia can alter gonadotropin secretion.

Surprisingly, in longstanding severe hypothyroidism, sexual precocity has been described too.[109] It has been hypothesized that this kind of "pseudopuberty," characterized by normal levels of gonadotropins, is due to a cross-interaction of the extremely elevated serum TSH with the FSH receptor.[110]

Vice versa, in hyperthyroid children, acceleration in linear growth may occur, while puberty may be delayed. If menarche has already occurred, secondary amenorrhea can be experienced.[109]

During the fertile life of females, TH have a role in the regulation of the menstrual cycle. In fact, hypothyroidism in adult women can alter the pulsatility of LH secretion, causing ovulatory disorders, galactorrhea, hirsutism, and irregular menstrual cycles.[111]

Moreover TH receptors are expressed in cumulus cells, oocytes, and granulosa cells[112] and TH have been found in human follicular fluid.[113] Thus, TH could be involved in maturation of the preovulatory follicle. The influence of TH on ovarian follicles, in concert with gonadotropins, is evidenced by the development of polycystic ovaries in hypothyroid women[114] and the reduction of ovarian cyst size and number after treatment with levothyroxine.[115] Moreover, women with hypothyroidism have significantly lower possibility of achieving pregnancy after in vitro fertilization compared to euthyroid patients[116] and positivity of antithyroid antibodies per se adversely affects conception and pregnancy outcomes.[117] Anyway, screening of thyroid function in women planning pregnancy is still highly controversial. The most shared approach is to test only subjects with a personal history of thyroid disease or other associated medical conditions.

Another evidence of the interaction between gonadotropin and the thyroid axis is represented by the alterations of thyroid function after controlled ovarian hyperstimulation. During treatment and for 3 months after treatment, TSH increases significantly both in euthyroid and hypothyroid women.[118,119] A hypothetical mechanism could be that hyperstimulation leads to supraphysiologic estrogen levels, which can alter thyroid homeostasis. In fact, elevated estrogens increase the level of thyroxine binding globuline (TBG), causing a decrease in unbound serum TH. Consequently, there is an increased secretion of TSH from the pituitary. Anyway, considering the evidence of interaction between TSH and gonadotropin receptors, other molecular mechanisms should be responsible for thyroid alterations after ovarian hyperstimulation. It is evident that the characterization of thyroid function in hyperstimulated women is important because untreated hypothyroidism can lead to adverse pregnancy outcomes and may affect both the pregnancy rate and the health of the developing fetus.

During pregnancy, an even more evident dialog between TH and gonadotropins is established. Syncytiotrophoblast cells secrete increasing levels of hCG, which are capable of binding and activating the TSHR in the thyroid, stimulating the gland function. Moreover, hCG regulates the expression of the sodium-iodide symporter in syncytiotrophoblasts, where it carries out iodide transport from maternal to fetal circulation.[120] This peculiar effect of hCG compensates the reduction of unbound active TH, secondary to the estrogen-stimulated increase in TBG, ensuring adequate levels of TH. In fact, TH play an important role during implantation and the early stage of embryo development.[121] The peripheral metabolism of TH is modified too, because of DIO type 3 overexpression in placenta.[122] This enzyme catalyzes deiodination of the inner ring of TH, causing their inactivation. Maybe as a compensatory mechanism, the expression of TH transporters in placental cells increases with gestation.[123] Thus, it is a delicate and very articulate balance. The great modification of thyroid axis in pregnancy and the interference of hCG reveals a relevant role of TH in this physiological condition.

In conclusion, there is clinical evidence that TH have a role in regulating gonadal development, ovarian function, and pregnancy outcome. However, the nature of the interaction between pituitary-thyroid axis and gonadotropins, albeit suggested by clinical evidence, is not fully clarified in detail yet.

6.2 Gonadotropins and Insulin-Like Growth Factors

Insulin-like growth factors (IGFs) are usually studied only for their role as peripheral hormone in the somatotropic axis. Surprisingly, in the last few years they have been reconsidered for their role in ovarian function regulation.

They have earned the appellation of "cogonadotropins," by virtue of their synergistic actions with LH and FSH. The influence of IGFs on gonadal axis starts at the central level, since they play a role in the regulation of neuroendocrine functions, including direct actions on GnRH-secreting neurons.[124] In mice, the presence of insulin-like growth factor-1 (IGF-1) receptor, at both mRNA and protein level, was demonstrated in brain regions expressing GnRH.[125] Moreover, IGF-1 receptor is expressed in the perikarya of GnRH-secreting neurons, suggesting a potential direct anatomical locus where IGF-1 can regulate reproductive development and function.[125]

Another central mechanism of interference between somatotropic and gonadotropic axes is represented by the human growth hormone (GH) binding to both GH and PRL receptors, leading to both somatotropic and lactotropic effects.[126] Thus, through the inhibitory effect of PRL, GH should indirectly alter gonadotropin secretion from the pituitary. In fact, in humans, excessive production of GH in acromegaly and GH resistance in Laron syndrome adversely affect reproduction. Similarly, in IGF-1 gene-knockout mice, the onset of puberty and/or fertility is altered.[127] Considering the IGFs' family, it is important to underline that the synthesis and actions of IGF-binding proteins (IGFBPs) are controlled by IGF-1.[128] In turn, binding proteins regulate IGF-1 action. Therefore, the concentration of IGFBP has a role in modulating IGF-1 effects on the neuroendocrine-gonadal system.

The influence of IGFs on the gonadal axis is exerted at the ovarian level too. In vitro studies with human thecal and granulosa cells demonstrated a key intraovarian role in follicle growth, selection, atresia, cellular differentiation and steroidogenesis, oocyte maturation, and cumulus expansion.[129] Interestingly, IGF-1 is required for the differentiation of preantral to preovulatory granulosa cells, in terms of response to FSH, via an AKT-mediated mechanism.[130] Moreover, IGFs potentiate the action of gonadotropins on ovarian steroid synthesis.[127]

Finally, IGF-1 receptors are expressed in the endometrium. They mediate cell proliferation and are upregulated in the secretory phase of the natural cycle, as a result of progesterone influence.[131] Insulin-like growth factor-2 promotes proliferation in the secretory phase and its action is more important during pregnancy, enhancing the proteolytic activity in placentation.[131]

6.3 Molecular Basis of Hormone Interactions

The interaction between gonadotropins and other hormones is easily comprehensible when the relationship involves glycoprotein hormones. While FSH and TSH bind to different and specific receptors, LH and CG share the same receptor. Glycoprotein hormones have the same heterodimeric structure composed of noncovalently linked α- and β-subunits. The α-subunit is identical within a species, while the β-subunit is hormone-specific.[132] Subunits also have a different role in regulating the interaction with receptors: the analysis of interactions in FSH–FSHR and TSH–TSHR complexes demonstrated that α-chains are involved in receptor activation process, while the β-chains define binding specificity.[133]

The newest member of the glycoprotein hormone family is the corticotroph-derived glycoprotein hormone (CGH), also referred to as thyrostimulin.[134] It represents a remarkable example of interaction between pituitary hormones, since it is expressed in the corticotrope cells of anterior pituitary, as well as in pancreas, skin, retina, and testis and activates the TSHR.[134] CGH stimulates cAMP production and thymidine incorporation in thyroid cells in vitro and stimulates thyroxine production in TSH-suppressed rats.[134] The role of thyrostimulin in human thyroid physiology, however, is still unknown. Surprisingly, this heterodimer is expressed in the follicle, where it can, as a paracrine factor, bind to TSHR and activate cAMP and the *c-fos* nuclear cascade in granulosa cells.[135] In turn, the ovarian expression of TSHR is clearly demonstrated[121] and tightly regulated by gonadotropins.[135] These findings confirm the importance of TSHR signaling in ovarian physiology and suggest that a paracrine thyrostimulin-based system, interacting with gonadotropins, is present in the ovary.

In addition to having a structural homology, glycoprotein hormones bind to structurally similar receptors. These receptors belong to the rhodopsin-like class of G protein-coupled receptors and are ~45% identical at the amino-acid level. The ancient precursors of glycoprotein hormones had a more promiscuous interaction behavior, which gradually diminished during evolution, resulting in increased hormone-receptor specificity.[136]

Also the known interaction between hCG and TSHR is explained by the structural homology not only in the hCG and TSH molecules but also in their receptors.[137] Actually, LH is able to bind the TSHR and is about 10 times more potent than hCG in increasing cAMP,[138] but the physiological LH levels found in vivo are not able to exert such a thyrotropic action. The very high levels reached by hCG in early pregnancy result in interaction with the TSHR, in spite of a lower potency compared to LH. If hCG were as potent as LH in regard to its thyrotropic activity, most pregnant women would become thyrotoxic. The difference between LH and hCG behavior, in spite of their high amino-acid sequence identity, is due to the CTP extension that is present only in hCG.[28] This sequence seems to prevent overt hyperthyroidism in the first trimester of pregnancy when a large amount of hCG is produced by the placenta. Lastly, the thyrotropic activity of hCG is influenced by the metabolic state of the hCG molecule itself. Deglycosylation and/or desialylation of hCG enhances its thyrotropic potency,[137] as well as partial digestion with carboxypeptidase.[139]

In spite of the described TSHR-IGF-1 receptor cross-talk,[140] especially within the context of thyroid-related pathology, such as Grave's ophthalmopathy,[141] no direct molecular cross-interactions between IGF-1 and the gonadotropin receptors were found.[142] Taken together, this suggests that the action of IGFs and all glycoprotein hormones are exerted in tandem by intracellular cross-talk, rather than by cross-interaction at the receptor level.

6.4 Gonadotropins and Cancer

Hormones play a physiological role in cell proliferation, differentiation, and apoptosis. For this reason, studying their role in cancer genesis and progression appears fascinating. Moreover, several tumors express hormones, or their receptors, at high levels, again suggesting a possible role in cancer progression.[143] In particular, gonadotropins are still widely studied for their role in the maintenance of the proper balance between cell proliferation and differentiation, and therefore in cancer.

FSH and LH have been implicated in a variety of different aspects of ovarian cancer tumorigenesis, including cellular proliferation, migration, and invasion.[144] There is some evidence that FSH and LH exert different actions on their target cells depending on the degree of differentiation: they stimulate proliferation when the cells are in an undifferentiated state and regulate only functional parameters when the cells are fully differentiated.[143] Considering this, it seems possible that some undifferentiated cancers produce gonadotropins in order to promote their own growth.[143]

Both FSH and LH actions are mediated mainly by cAMP/PKA signaling (Fig. 2). Notably, the activity of the ERK signaling cascade is elevated in response to these stimuli as well. ERK signaling pathway is known to promote cell survival by a dual mechanism comprising the posttranslational inactivation of proapoptotic factors and the increased transcription of prosurvival genes.[145] However, in granulosa cells, the FSH-mediated steroidogenic effects in vitro are mediated mainly by cAMP, especially at high doses, and result in proapoptotic consequences.[146,147]

A separate discussion should be reserved to hCG. The hyperglycosylated form of hCG comprises up to 90% of total hCG measurable in serum and urine during the first 3 weeks of pregnancy when invasive trophoblast activity is high.[148] Besides promoting trophoblast invasion, it also has a potential role in immune cell modulation within the uterus at the time of pregnancy initiation.[148] Obviously, both of these features make it a special candidate as a regulator of cancer genesis and growth. Indeed, hCG is expressed by both trophoblastic and nontrophoblastic human malignancies and is a clinical marker of placental trophoblastic malignancies and germ cell tumors of the testis and the ovary.[149]

7. CONCLUSIONS

Gonadotropins belong to a heterogeneous family comprising various molecules that differ at several levels. First of all, they are encoded by specific genes that underwent a strong selection during mammalian evolution, revealing the endocrine adaptation to reproduction. This complex mechanism is reflected in specific effects of each gonadotropin at the molecular level. However, the commercially available gonadotropins are not able to fully reproduce the wide complexity of the endogenous counterparts. Therefore, different protocols are used in ART and personalized therapy is empirically used in clinical practice in order to avoid adverse events. Finally, there is clear evidence of a tight interaction between gonadotropins and other endocrine axes that synergistically regulate reproduction, modulating the proliferation/differentiation balance in target tissues. Taken together, all these issues reveal the refined system in which gonadotropins are the main actors of reproduction, the most important biological function for the maintenance of human species.

ACKNOWLEDGMENTS

LC was supported by a grant of the Italian Ministry of Education, University and Research (PRIN 2010C8ERKX) and by the Associazione Scientifica in Endocrinologia, Andrologia e Metabolismo (ASEAM), Carpi, Italy. GB and DS are fellows of the PhD School in Clinical and Experimental Medicine of the University of Modena and Reggio Emilia, Italy.

REFERENCES

1. Ezcurra D, Humaidan P. A review of luteinising hormone and human chorionic gonadotropin when used in assisted reproductive technology. *Reprod Biol Endocrinol.* 2014;12:95.
2. Hillier SG. Gonadotropic control of ovarian follicular growth and development. *Mol Cell Endocrinol.* 2001;179(1–2):39–46.

3. Usadi RS, Merriam KS. On-label and off-label drug use in the treatment of female infertility. *Fertil Steril.* 2015;103(3):583–594.
4. Wolfenson C, Groisman J, Couto AS, et al. Batch-to-batch consistency of human-derived gonadotrophin preparations compared with recombinant preparations. *Reprod Biomed Online.* 2005;10(4):442–454.
5. Casarini L, Santi D, Marino M. Impact of gene polymorphisms of gonadotropins and their receptors on human reproductive success. *Reproduction.* 2015;150(6): R175–R184.
6. Eastwick PW, Finkel EJ. The evolutionary armistice: attachment bonds moderate the function of ovulatory cycle adaptations. *Pers Soc Psychol Bull.* 2012;38(2):174–184.
7. Vitzthum VJ. The ecology and evolutionary endocrinology of reproduction in the human female. *Am J Phys Anthropol.* 2009;140(suppl 49):95–136.
8. Huhtaniemi I. A short evolutionary history of FSH-stimulated spermatogenesis. *Hormones.* 2015;14(4):468–478.
9. Schlatt S, Ehmcke J. Regulation of spermatogenesis: an evolutionary biologist's perspective. *Semin Cell Dev Biol.* 2014;29:2–16.
10. Civetta A. Adaptive evolution at immune system genes and deep pregnancy implantation in primates. *Genomics.* 2015;105(1):17–22.
11. Cole LA. hCG and hyperglycosylated hCG in the establishment and evolution of hemochorial placentation. *J Reprod Immunol.* 2009;82(2):112–118.
12. Smits G, Campillo M, Govaerts C, et al. Glycoprotein hormone receptors: determinants in leucine-rich repeats responsible for ligand specificity. *EMBO J.* 2003;22 (11):2692–2703.
13. Ascoli M, Fanelli F, Segaloff DL. The lutropin/choriogonadotropin receptor, a 2002 perspective. *Endocr Rev.* 2002;23(2):141–174.
14. Plunkett J, Doniger S, Orabona G, et al. An evolutionary genomic approach to identify genes involved in human birth timing. *PLoS Genet.* 2011;7(4):e1001365.
15. Behre HM, Greb RR, Mempel A, et al. Significance of a common single nucleotide polymorphism in exon 10 of the follicle-stimulating hormone (FSH) receptor gene for the ovarian response to FSH: a pharmacogenetic approach to controlled ovarian hyperstimulation. *Pharmacogenet Genomics.* 2005;15(7):451–456.
16. Simoni M, Casarini L. Mechanisms in endocrinology: genetics of FSH action: a 2014-and-beyond view. *Euro J Endocrinol.* 2014;170(3):R91–R107.
17. Tuttelmann F, Laan M, Grigorova M, Punab M, Sober S, Gromoll J. Combined effects of the variants FSHB -211G>T and FSHR 2039A>G on male reproductive parameters. *J Clin Endocrinol Metab.* 2012;97(10):3639–3647.
18. Hallast P, Rull K, Laan M. The evolution and genomic landscape of CGB1 and CGB2 genes. *Mol Cell Endocrinol.* 2007;260–262:2–11.
19. Gromoll J, Eiholzer U, Nieschlag E, Simoni M. Male hypogonadism caused by homozygous deletion of exon 10 of the luteinizing hormone (LH) receptor: differential action of human chorionic gonadotropin and LH. *J Clin Endocrinol Metab.* 2000;85(6): 2281–2286.
20. Muller T, Simoni M, Pekel E, et al. Chorionic gonadotrophin beta subunit mRNA but not luteinising hormone beta subunit mRNA is expressed in the pituitary of the common marmoset (*Callithrix jacchus*). *J Mol Endocrinol.* 2004;32(1):115–128.
21. Perez Mayorga M, Gromoll J, Behre HM, Gassner C, Nieschlag E, Simoni M. Ovarian response to follicle-stimulating hormone (FSH) stimulation depends on the FSH receptor genotype. *J Clin Endocrinol Metab.* 2000;85(9):3365–3369.
22. Simoni M, Gromoll J, Hoppner W, et al. Mutational analysis of the follicle-stimulating hormone (FSH) receptor in normal and infertile men: identification and characterization of two discrete FSH receptor isoforms. *J Clin Endocrinol Metab.* 1999;84(2): 751–755.

23. Sudo S, Kudo M, Wada S, Sato O, Hsueh AJ, Fujimoto S. Genetic and functional analyses of polymorphisms in the human FSH receptor gene. *Mol Hum Reprod.* 2002;8(10):893–899.
24. Ryan GL, Feng X, d'Alva CB, et al. Evaluating the roles of follicle-stimulating hormone receptor polymorphisms in gonadal hyperstimulation associated with severe juvenile primary hypothyroidism. *J Clin Endocrinol Metab.* 2007;92(6):2312–2317.
25. Nordhoff V, Sonntag B, von Tils D, et al. Effects of the FSH receptor gene polymorphism p.N680S on cAMP and steroid production in cultured primary human granulosa cells. *Reprod Biomed Online.* 2011;23(2):196–203.
26. Rich TC, Xin W, Mehats C, et al. Cellular mechanisms underlying prostaglandin-induced transient cAMP signals near the plasma membrane of HEK-293 cells. *Am J Physiol Cell Physiol.* 2007;292(1):C319–C331.
27. Casarini L, Moriondo V, Marino M, et al. FSHR polymorphism p.N680S mediates different responses to FSH in vitro. *Mol Cell Endocrinol.* 2014;393(1–2):83–91.
28. Choi J, Smitz J. Luteinizing hormone and human chorionic gonadotropin: origins of difference. *Mol Cell Endocrinol.* 2014;383(1–2):203–213.
29. Muller T, Gromoll J, Simoni M. Absence of exon 10 of the human luteinizing hormone (LH) receptor impairs LH, but not human chorionic gonadotropin action. *J Clin Endocrinol Metab.* 2003;88(5):2242–2249.
30. Grzesik P, Kreuchwig A, Rutz C, et al. Differences in signal activation by LH and hCG are mediated by the LH/CG receptor's extracellular Hinge region. *Front Endocrinol.* 2015;6:140.
31. Landomiel F, Gallay N, Jegot G, et al. Biased signalling in follicle stimulating hormone action. *Mol Cell Endocrinol.* 2014;382(1):452–459.
32. Casarini L, Lispi M, Longobardi S, et al. LH and hCG action on the same receptor results in quantitatively and qualitatively different intracellular signalling. *PloS One.* 2012;7(10):e46682.
33. Casarini L, Riccetti L, De Pascali F, et al. Follicle-stimulating hormone potentiates the steroidogenic activity of chorionic gonadotropin and the anti-apoptotic activity of luteinizing hormone in human granulosa-lutein cells in vitro. *Mol Cell Endocrinol.* 2016;422:103–114.
34. Menon KM, Menon B. Regulation of luteinizing hormone receptor expression by an RNA binding protein: role of ERK signaling. *Indian J Med Res.* 2014;140(suppl):S112–S119.
35. Noma N, Kawashima I, Fan HY, et al. LH-induced neuregulin 1 (NRG1) type III transcripts control granulosa cell differentiation and oocyte maturation. *Mol Endocrinol.* 2011;25(1):104–116.
36. Gupta C, Chapekar T, Chhabra Y, Singh P, Sinha S, Luthra K. Differential response to sustained stimulation by hCG & LH on goat ovarian granulosa cells. *Indian J Med Res.* 2012;135:331–340.
37. Amsterdam A, Gold RS, Hosokawa K, et al. Crosstalk among multiple signaling pathways controlling ovarian cell death. *Trends Endocrinol Metab.* 1999;10(7):255–262.
38. Lehert P, Schertz JC, Ezcurra D. Recombinant human follicle-stimulating hormone produces more oocytes with a lower total dose per cycle in assisted reproductive technologies compared with highly purified human menopausal gonadotrophin: a meta-analysis. *Reprod Biol Endocrinol.* 2010;8:112.
39. Santi D, Simoni M. Biosimilar recombinant follicle stimulating hormones in infertility treatment. *Expert Opin Biol Ther.* 2014;14(10):1399–1409.
40. Bassett RM, Driebergen R. Continued improvements in the quality and consistency of follitropin alfa, recombinant human FSH. *Reprod Biomed Online.* 2005;10(2):169–177.
41. Keene JL, Matzuk MM, Otani T, et al. Expression of biologically active human follitropin in Chinese hamster ovary cells. *J Biol Chem.* 1989;264(9):4769–4775.

42. van de Weijer BH, Mulders JW, Bos ES, Verhaert PD, van den Hooven HW. Compositional analyses of a human menopausal gonadotrophin preparation extracted from urine (menotropin). Identification of some of its major impurities. *Reprod Biomed Online.* 2003;7(5):547–557.
43. Giudice E, Crisci C, Eshkol A, Papoian R. Composition of commercial gonadotrophin preparations extracted from human post-menopausal urine: characterization of non-gonadotrophin proteins. *Hum Reprod.* 1994;9(12):2291–2299.
44. Yarram SJ, Jenkins J, Cole LA, Brown NL, Sandy JR, Mansell JP. Epidermal growth factor contamination and concentrations of intact human chorionic gonadotropin in commercial preparations. *Fertil Steril.* 2004;82(1):232–233.
45. Kuwabara Y, Mine K, Katayama A, Inagawa T, Akira S, Takeshita T. Proteomic analyses of recombinant human follicle-stimulating hormone and urinary-derived gonadotropin preparations. *J Reprod Med.* 2009;54(8):459–466.
46. Van Dorsselaer A, Carapito C, Delalande F, et al. Detection of prion protein in urine-derived injectable fertility products by a targeted proteomic approach. *PloS One.* 2011;6 (3):e17815.
47. Imran M, Mahmood S. An overview of human prion diseases. *Virol J.* 2011;8:559.
48. Ward HJ, Balen A, Will RG. Creutzfeldt–Jakob disease and urinary gonadotrophins. *Hum Reprod.* 2004;19(5):1236–1237.
49. Rose MP, Gaines Das RE, Balen AH. Definition and measurement of follicle stimulating hormone. *Endocr Rev.* 2000;21(1):5–22.
50. Howles CM, Loumaye E, Giroud D, Luyet G. Multiple follicular development and ovarian steroidogenesis following subcutaneous administration of a highly purified urinary FSH preparation in pituitary desensitized women undergoing IVF: a multicentre European phase III study. *Hum Reprod.* 1994;9(3):424–430.
51. Howles CM. Genetic engineering of human FSH (Gonal-F). *Hum Reprod Update.* 1996;2(2):172–191.
52. Bergh C. What are the clinical benefits of recombinant gonadotrophins?. Recombinant follicle stimulating hormone. *Hum Reprod.* 1999;14(6):1418–1420.
53. Lunenfeld B. Development of gonadotrophins for clinical use. *Reprod Biomed Online.* 2002;4(suppl 1):11–17.
54. Butnev VY, Butnev VY, May JV, et al. Production, purification, and characterization of recombinant hFSH glycoforms for functional studies. *Mol Cell Endocrinol.* 2015;405:42–51.
55. Jiang C, Hou X, Wang C, et al. Hypoglycosylated hFSH has greater bioactivity than fully glycosylated recombinant hFSH in human granulosa cells. *J Clin Endocrinol Metab.* 2015;100(6):E852–E860.
56. de Leeuw R, Mulders J, Voortman G, Rombout F, Damm J, Kloosterboer L. Structure-function relationship of recombinant follicle stimulating hormone (Puregon). *Mol Hum Reprod.* 1996;2(5):361–369.
57. Hard K, Mekking A, Damm JB, et al. Isolation and structure determination of the intact sialylated N-linked carbohydrate chains of recombinant human follitropin expressed in Chinese hamster ovary cells. *Eur J Biochem/FEBS.* 1990;193(1):263–271.
58. Lambert A, Rodgers M, Mitchell R, et al. In-vitro biopotency and glycoform distribution of recombinant human follicle stimulating hormone (Org 32489), Metrodin and Metrodin-HP. *Hum Reprod.* 1995;10(7):1928–1935.
59. Rose TH, Roshammar D, Erichsen L, Grundemar L, Ottesen JT. Characterisation of population pharmacokinetics and endogenous follicle-stimulating hormone (FSH) levels after multiple dosing of a recombinant human FSH (FE 999049) in healthy women. *Drugs R D.* 2016;16(2):165–172.
60. Brinsden P, Akagbosu F, Gibbons LM, et al. A comparison of the efficacy and tolerability of two recombinant human follicle-stimulating hormone preparations in patients undergoing in vitro fertilization-embryo transfer. *Fertil Steril.* 2000;73(1):114–116.

61. de Lartigue J. Corifollitropin alfa: a new option to treat female infertility. *Drugs Today (Barc).* 2011;47(8):583–590.
62. Roger SD, Mikhail A. Biosimilars: opportunity or cause for concern? *J Pharm Pharm Sci.* 2007;10(3):405–410.
63. Baker VL, Brown MB, Luke B, Conrad KP. Association of number of retrieved oocytes with live birth rate and birth weight: an analysis of 231,815 cycles of in vitro fertilization. *Fertil Steril.* 2015;103(4). 931–938 e932.
64. Lombardi A, Andreozzi C, Pavone V, Triglione V, Angiolini L, Caccia P. Evaluation of the oligosaccharide composition of commercial follicle stimulating hormone preparations. *Electrophoresis.* 2013;34(16):2394–2406.
65. Cantineau AE, Cohlen BJ, Heineman MJ. Ovarian stimulation protocols (anti-oestrogens, gonadotrophins with and without GnRH agonists/antagonists) for intrauterine insemination (IUI) in women with subfertility. *Cochrane Database Syst Rev.* 2007;2: CD005356.
66. van Wely M, Kwan I, Burt AL, et al. Recombinant versus urinary gonadotrophin for ovarian stimulation in assisted reproductive technology cycles. *Cochrane Database Syst Rev.* 2011;2:CD005354.
67. Alviggi C, Humaidan P, Ezcurra D. Hormonal, functional and genetic biomarkers in controlled ovarian stimulation: tools for matching patients and protocols. *Reprod Biol Endocrinol.* 2012;10:9.
68. La Marca A, Ferraretti AP, Palermo R, Ubaldi FM. The use of ovarian reserve markers in IVF clinical practice: a national consensus. *Gynecol Endocrinol.* 2016;32(1):1–5.
69. Siristatidis CS, Gibreel A, Basios G, Maheshwari A, Bhattacharya S. Gonadotrophin-releasing hormone agonist protocols for pituitary suppression in assisted reproduction. *Cochrane Database Syst Rev.* 2015;(11):CD006919.
70. Copperman AB, Benadiva C. Optimal usage of the GnRH antagonists: a review of the literature. *Reprod Biol Endocrinol.* 2013;11:20.
71. Custers IM, Steures P, van der Steeg JW, et al. External validation of a prediction model for an ongoing pregnancy after intrauterine insemination. *Fertil Steril.* 2007;88 (2):425–431.
72. Imani B, Eijkemans MJ, Faessen GH, Bouchard P, Giudice LC, Fauser BC. Prediction of the individual follicle-stimulating hormone threshold for gonadotropin induction of ovulation in normogonadotropic anovulatory infertility: an approach to increase safety and efficiency. *Fertil Steril.* 2002;77(1):83–90.
73. Steures P, van der Steeg JW, Mol BW, et al. Prediction of an ongoing pregnancy after intrauterine insemination. *Fertil Steril.* 2004;82(1):45–51.
74. Daya S, Gunby J. WITHDRAWN: recombinant versus urinary follicle stimulating hormone for ovarian stimulation in assisted reproduction cycles. *Cochrane Database Syst Rev.* 2006;3:CD002810.
75. Gholami H, Vicari E, Molis M, La Vignera S, Papaleo E, Cappiello F. Pregnancy outcome following in vitro fertilization-embryo transfer (IVF-ET) in women aged < 37, undergoing ovulation induction with human FSH compared with recombinant FSH: a randomised controlled study. *Eur Rev Med Pharmacol Sci.* 2010;14(2): 97–102.
76. Pouwer AW, Farquhar C, Kremer JA, Long-acting FSH. versus daily FSH for women undergoing assisted reproduction. *Cochrane Database Syst Rev.* 2015;7:CD009577.
77. Olivennes F, Howies CM, Borini A, et al. Individualizing FSH dose for assisted reproduction using a novel algorithm: the CONSORT study. *Reprod Biomed Online.* 2011;22 (suppl 1):S73–S82.
78. Olivennes F, Trew G, Borini A, et al. Randomized, controlled, open-label, non-inferiority study of the CONSORT algorithm for individualized dosing of follitropin alfa. *Reprod Biomed Online.* 2015;30(3):248–257.

79. Pouly JL, Olivennes F, Massin N, Celle M, Caizergues N, Contard F. Usability and utility of the CONSORT calculator for FSH starting doses: a prospective observational study. *Reprod Biomed Online.* 2015;31(3):347–355.
80. Simon C, Garcia Velasco JJ, Valbuena D, et al. Increasing uterine receptivity by decreasing estradiol levels during the preimplantation period in high responders with the use of a follicle-stimulating hormone step-down regimen. *Fertil Steril.* 1998;70 (2):234–239.
81. Tavaniotou A, Albano C, Smitz J, Devroey P. Impact of ovarian stimulation on corpus luteum function and embryonic implantation. *J Reprod Immunol.* 2002;55(1–2): 123–130.
82. Tay CC. Use of gonadotrophin-releasing hormone agonists to trigger ovulation. *Hum Fertil (Camb).* 2002;5(1):G35–G37. discussion G38–39, G41–38.
83. Youssef MA, Van der Veen F, Al-Inany HG, et al. Gonadotropin-releasing hormone agonist versus HCG for oocyte triggering in antagonist-assisted reproductive technology. *Cochrane Database Syst Rev.* 2014;10:CD008046.
84. Baker JW, Morgan RL, Peckham MJ, Smithers DW. Preservation of ovarian function in patients requiring radiotherapy for para-aortic and pelvic Hodgkin's disease. *Lancet.* 1972;1(7764):1307–1308.
85. Hill MJ, Levy G, Levens ED. Does exogenous LH in ovarian stimulation improve assisted reproduction success? An appraisal of the literature. *Reprod Biomed Online.* 2012;24(3):261–271.
86. Lehert P, Kolibianakis EM, Venetis CA, et al. Recombinant human follicle-stimulating hormone (r-hFSH) plus recombinant luteinizing hormone versus r-hFSH alone for ovarian stimulation during assisted reproductive technology: systematic review and meta-analysis. *Reprod Biol Endocrinol.* 2014;12:17.
87. Azziz R. Introduction: determinants of polycystic ovary syndrome. *Fertil Steril.* 2016;106:4–5.
88. Goodman NF, Cobin RH, Futterweit W, Glueck JS, Legro RS, Carmina E. American Association of Clinical Endocrinologists, American College of Endocrinology, and Androgen Excess and PCOS Society Disease State Clinical Review: guide to the best practices in the evaluation and treatment of polycystic ovary syndrome—Part 1. *Endocr Pract.* 2015;21(11):1291–1300.
89. Goodman NF, Cobin RH, Futterweit W, Glueck JS, Legro RS, Carmina E. American Association of Clinical Endocrinologists, American College of Endocrinology, and Androgen Excess and PCOS Society Disease State Clinical Review: guide to the best practices in the evaluation and treatment of polycystic ovary syndrome—Part 2. *Endocr Pract.* 2015;21(12):1415–1426.
90. La Marca A, Sunkara SK. Individualization of controlled ovarian stimulation in IVF using ovarian reserve markers: from theory to practice. *Hum Reprod Update.* 2014;20 (1):124–140.
91. Smith V, Osianlis T, Vollenhoven B. Prevention of ovarian hyperstimulation syndrome: a review. *Obstet Gynecol Int.* 2015;2015:514159.
92. Nastri CO, Ferriani RA, Rocha IA, Martins WP. Ovarian hyperstimulation syndrome: pathophysiology and prevention. *J Assist Reprod Genet.* 2010;27(2–3):121–128.
93. Dieterich M, Bolz M, Reimer T, Costagliola S, Gerber B. Two different entities of spontaneous ovarian hyperstimulation in a woman with FSH receptor mutation. *Reprod Biomed Online.* 2010;20(6):751–758.
94. Rizk B. Symposium: update on prediction and management of OHSS. Genetics of ovarian hyperstimulation syndrome. *Reprod Biomed Online.* 2009;19(1):14–27.
95. Wang TH, Horng SG, Chang CL, et al. Human chorionic gonadotropin-induced ovarian hyperstimulation syndrome is associated with up-regulation of vascular endothelial growth factor. *J Clin Endocrinol Metab.* 2002;87(7):3300–3308.

96. Kasum M. New insights in mechanisms for development of ovarian hyperstimulation syndrome. *Coll Antropol.* 2010;34(3):1139–1143.
97. Goldsman MP, Pedram A, Dominguez CE, Ciuffardi I, Levin E, Asch RH. Increased capillary permeability induced by human follicular fluid: a hypothesis for an ovarian origin of the hyperstimulation syndrome. *Fertil Steril.* 1995;63(2):268–272.
98. Boothroyd C, Karia S, Andreadis N, Rombauts L, Johnson N, Chapman M. Consensus statement on prevention and detection of ovarian hyperstimulation syndrome. *Aust New Zealand J Obstet Gynaecol.* 2015;55(6):523–534.
99. Faddy MJ, Gosden RG, Gougeon A, Richardson SJ, Nelson JF. Accelerated disappearance of ovarian follicles in mid-life: implications for forecasting menopause. *Hum Reprod.* 1992;7(10):1342–1346.
100. Seifer DB, Scott Jr RT, Bergh PA, et al. Women with declining ovarian reserve may demonstrate a decrease in day 3 serum inhibin B before a rise in day 3 follicle-stimulating hormone. *Fertil Steril.* 1999;72(1):63–65.
101. Volarcik K, Sheean L, Goldfarb J, Woods L, Abdul-Karim FW, Hunt P. The meiotic competence of in-vitro matured human oocytes is influenced by donor age: evidence that folliculogenesis is compromised in the reproductively aged ovary. *Hum Reprod.* 1998;13(1):154–160.
102. Ferraretti AP, La Marca A, Fauser BC, Tarlatzis B, Nargund G, Gianaroli L. ESHRE consensus on the definition of 'poor response' to ovarian stimulation for in vitro fertilization: the Bologna criteria. *Hum Reprod.* 2011;26(7):1616–1624.
103. Farhi J, Jacobs HS. Early prediction of ovarian multifollicular response during ovulation induction in patients with polycystic ovary syndrome. *Fertil Steril.* 1997;67(3): 459–462.
104. Nikolaou D, Lavery S, Turner C, Margara R, Trew G. Is there a link between an extremely poor response to ovarian hyperstimulation and early ovarian failure? *Hum Reprod.* 2002;17(4):1106–1111.
105. Zhao J, Li Y, Zhang Q, Wang Y. Does ovarian stimulation for IVF increase gynaecological cancer risk?. A systematic review and meta-analysis. *Reprod Biomed Online.* 2015;31(1):20–29.
106. Davis PJ, Leonard JL, Davis FB. Mechanisms of nongenomic actions of thyroid hormone. *Front Neuroendocrinol.* 2008;29(2):211–218.
107. Flood DE, Fernandino JI, Langlois VS. Thyroid hormones in male reproductive development: evidence for direct crosstalk between the androgen and thyroid hormone axes. *Gen Comp Endocrinol.* 2013;192:2–14.
108. Kakar SS, Nath S, Bunn J, Jennes L. The inhibition of growth and down-regulation of gonadotropin releasing hormone (GnRH) receptor in alphaT3-1 cells by GnRH agonist. *Anticancer Drugs.* 1997;8(4):369–375.
109. Weber G, Vigone MC, Stroppa L, Chiumello G. Thyroid function and puberty. *J Pediatr Endocrinol Metab.* 2003;16(suppl 2):253–257.
110. Anasti JN, Flack MR, Froehlich J, Nelson LM, Nisula BC. A potential novel mechanism for precocious puberty in juvenile hypothyroidism. *J Clin Endocrinol Metab.* 1995;80(1):276–279.
111. Negro R, Soldin OP, Obregon MJ, Stagnaro-Green A. Hypothyroxinemia and pregnancy. *Endocr Pract.* 2011;17(3):422–429.
112. Poppe K, Velkeniers B. Female infertility and the thyroid. *Best Pract Res Clin Endocrinol Metab.* 2004;18(2):153–165.
113. Wakim AN, Polizotto SL, Buffo MJ, Marrero MA, Burholt DR. Thyroid hormones in human follicular fluid and thyroid hormone receptors in human granulosa cells. *Fertil Steril.* 1993;59(6):1187–1190.
114. Du D, Li X. The relationship between thyroiditis and polycystic ovary syndrome: a meta-analysis. *Int J Clin Exp Med.* 2013;6(10):880–889.

115. Lindsay AN, Voorhess ML, MacGillivray MH. Multicystic ovaries in primary hypothyroidism. *Obstet Gynecol.* 1983;61(4):433–437.
116. Scoccia B, Demir H, Kang Y, Fierro MA, Winston NJ. In vitro fertilization pregnancy rates in levothyroxine-treated women with hypothyroidism compared to women without thyroid dysfunction disorders. *Thyroid.* 2012;22(6):631–636.
117. van den Boogaard E, Vissenberg R, Land JA, et al. Significance of (sub)clinical thyroid dysfunction and thyroid autoimmunity before conception and in early pregnancy: a systematic review. *Hum Reprod Update.* 2011;17(5):605–619.
118. Busnelli A, Somigliana E, Ferrari S, et al. The long-term impact of controlled ovarian hyperstimulation on thyroid function. *Endocr Pract.* 2016;22(4):389–395.
119. Gracia CR, Morse CB, Chan G, et al. Thyroid function during controlled ovarian hyperstimulation as part of in vitro fertilization. *Fertil Steril.* 2012;97(3):585–591.
120. Chan SY, Vasilopoulou E, Kilby MD. The role of the placenta in thyroid hormone delivery to the fetus. *Nat Clin Pract Endocrinol Metab.* 2009;5(1):45–54.
121. Colicchia M, Campagnolo L, Baldini E, Ulisse S, Valensise H, Moretti C. Molecular basis of thyrotropin and thyroid hormone action during implantation and early development. *Hum Reprod Update.* 2014;20(6):884–904.
122. Kurlak LO, Mistry HD, Kaptein E, Visser TJ, Broughton Pipkin F. Thyroid hormones and their placental deiodination in normal and pre-eclamptic pregnancy. *Placenta.* 2013;34(5):395–400.
123. Loubiere LS, Vasilopoulou E, Bulmer JN, et al. Expression of thyroid hormone transporters in the human placenta and changes associated with intrauterine growth restriction. *Placenta.* 2010;31(4):295–304.
124. Daftary SS, Gore AC. IGF-1 in the brain as a regulator of reproductive neuroendocrine function. *Exp Biol Med.* 2005;230(5):292–306.
125. Daftary SS, Gore AC. The hypothalamic insulin-like growth factor-1 receptor and its relationship to gonadotropin-releasing hormones neurones during postnatal development. *J Neuroendocrinol.* 2004;16(2):160–169.
126. Chandrashekar V, Bartke A. The role of insulin-like growth factor-I in neuroendocrine function and the consequent effects on sexual maturation: inferences from animal models. *Reprod Biol.* 2003;3(1):7–28.
127. Chandrashekar V, Zaczek D, Bartke A. The consequences of altered somatotropic system on reproduction. *Biol Reprod.* 2004;71(1):17–27.
128. Giudice LC, van Dessel HJ, Cataldo NA, Chandrasekher YA, Yap OW, Fauser BC. Circulating and ovarian IGF binding proteins: potential roles in normo-ovulatory cycles and in polycystic ovarian syndrome. *Prog Growth Factor Res.* 1995;6(2–4): 397–408.
129. Kwintkiewicz J, Giudice LC. The interplay of insulin-like growth factors, gonadotropins, and endocrine disruptors in ovarian follicular development and function. *Semin Reprod Med.* 2009;27(1):43–51.
130. Baumgarten SC, Convissar SM, Fierro MA, Winston NJ, Scoccia B, Stocco C. IGF1R signaling is necessary for FSH-induced activation of AKT and differentiation of human Cumulus granulosa cells. *J Clin Endocrinol Metab.* 2014;99(8):2995–3004.
131. Detti L, Uhlmann RA, Fletcher NM, Diamond MP, Saed GM. Endometrial signaling pathways during ovarian stimulation for assisted reproduction technology. *Fertil Steril.* 2013;100(3):889–894.
132. Pierce JG, Parsons TF. Glycoprotein hormones: structure and function. *Annu Rev Biochem.* 1981;50:465–495.
133. Nunez Miguel R, Sanders J, Chirgadze DY, Blundell TL, Furmaniak J, Rees Smith B. FSH and TSH binding to their respective receptors: similarities, differences and implication for glycoprotein hormone specificity. *J Mol Endocrinol.* 2008;41(3): 145–164.

134. Nakabayashi K, Matsumi H, Bhalla A, et al. Thyrostimulin, a heterodimer of two new human glycoprotein hormone subunits, activates the thyroid-stimulating hormone receptor. *J Clin Invest.* 2002;109(11):1445–1452.
135. Sun SC, Hsu PJ, Wu FJ, Li SH, Lu CH, Luo CW. Thyrostimulin, but not thyroid-stimulating hormone (TSH), acts as a paracrine regulator to activate the TSH receptor in mammalian ovary. *J Biol Chem.* 2010;285(6):3758–3765.
136. Moyle WR, Campbell RK, Myers RV, Bernard MP, Han Y, Wang X. Co-evolution of ligand-receptor pairs. *Nature.* 1994;368(6468):251–255.
137. Yoshimura M, Hershman JM. Thyrotropic action of human chorionic gonadotropin. *Thyroid.* 1995;5(5):425–434.
138. Yoshimura M, Hershman JM, Pang XP, Berg L, Pekary AE. Activation of the thyrotropin (TSH) receptor by human chorionic gonadotropin and luteinizing hormone in Chinese hamster ovary cells expressing functional human TSH receptors. *J Clin Endocrinol Metab.* 1993;77(4):1009–1013.
139. Carayon P, Amr S, Nisula B, Lissitzky S. Effect of carboxypeptidase digestion of the human choriogonadotropin molecule on its thyrotropic activity. *Endocrinology.* 1981;108(5):1891–1898.
140. Kohn LD, Shimura H, Shimura Y, et al. The thyrotropin receptor. *Vitam Horm.* 1995;50: 287–384.
141. Krieger CC, Neumann S, Place RF, Marcus-Samuels B, Gershengorn MC. Bidirectional TSH and IGF-1 receptor cross talk mediates stimulation of hyaluronan secretion by Graves' disease immunoglobins. *J Clin Endocrinol Metab.* 2015;100(3): 1071–1077.
142. Hsu SY, Liang SG, Hsueh AJ. Characterization of two LGR genes homologous to gonadotropin and thyrotropin receptors with extracellular leucine-rich repeats and a G protein-coupled, seven-transmembrane region. *Mol Endocrinol.* 1998;12(12): 1830–1845.
143. Govindaraj V, Arya SV, Rao AJ. Differential action of glycoprotein hormones: significance in cancer progression. *Horm Cancer.* 2014;5(1):1–10.
144. Mertens-Walker I, Baxter RC, Marsh DJ. Gonadotropin signalling in epithelial ovarian cancer. *Cancer Lett.* 2012;324(2):152–159.
145. Kolch W. Coordinating ERK/MAPK signalling through scaffolds and inhibitors. *Nat Rev Mol Cell Biol.* 2005;6(11):827–837.
146. Breckwoldt M, Selvaraj N, Aharoni D, et al. Expression of Ad4-BP/cytochrome P450 side chain cleavage enzyme and induction of cell death in long-term cultures of human granulosa cells. *Mol Hum Reprod.* 1996;2(6):391–400.
147. Schiffer D, Cavalla P, Migheli A, Giordana MT, Chiado-Piat L. Bcl-2 distribution in neuroepithelial tumors: an immunohistochemical study. *J Neurooncol.* 1996;27(2): 101–109.
148. Evans J. Hyperglycosylated hCG: a unique human implantation and invasion factor. *Am J Reprod Immunol.* 2016;75(3):333–340.
149. Sisinni L, Landriscina M. The role of human chorionic gonadotropin as tumor marker: biochemical and clinical aspects. *Adv Exp Med Biol.* 2015;867:159–176.

CHAPTER FOUR

Clinical Applications of Gonadotropins in the Male

A. Ulloa-Aguirre*[,1], S. Lira-Albarrán[†]

[*] Research Support Network, Universidad Nacional Autónoma de México (UNAM)-National Institutes of Health, Mexico City, Mexico

[†] Department of Reproductive Biology, Instituto Nacional de Ciencias Médicas y Nutrición Salvador Zubirán, Mexico City, Mexico

[1] Corresponding author. E-mail address: aulloaa@unam.mx

Contents

Abstract

The pituitary gonadotropins, luteinizing hormone (LH), and follicle-stimulating hormone (FSH) play a pivotal role in reproduction. The synthesis and secretion of gonadotropins are regulated by complex interactions among several endocrine, paracrine, and autocrine factors of diverse chemical structure. In men, LH regulates the synthesis of androgens by the Leydig cells, whereas FSH promotes Sertoli cell function and thereby influences spermatogenesis. Gonadotropins are complex molecules

Progress in Molecular Biology and Translational Science, Volume 143
ISSN 1877-1173
http://dx.doi.org/10.1016/bs.pmbts.2016.08.003

composed of two subunits, the α- and β-subunit, that are noncovalently associated. Gonadotropins are decorated with glycans that regulate several functions of the protein including folding, heterodimerization, stability, transport, conformational maturation, efficiency of heterodimer secretion, metabolic fate, interaction with their cognate receptor, and selective activation of signaling pathways. A number of congenital and acquired abnormalities lead to gonadotropin deficiency and hypogonadotropic hypogonadism, a condition amenable to treatment with exogenous gonadotropins. Several natural and recombinant preparations of gonadotropins are currently available for therapeutic purposes. The difference between natural and the currently available recombinant preparations (which are massively produced in Chinese hamster ovary cells for commercial purposes) mainly lies in the abundance of some of the carbohydrates that conform the complex glycans attached to the protein core. Whereas administration of exogenous gonadotropins in patients with isolated congenital hypogonadotropic hypogonadism is a well recognized therapeutic approach, their role in treating men with normogonadotropic idiopathic infertility is still controversial. This chapter concentrates on the main structural and functional features of the gonadotropin hormones and how basic concepts have been translated into the clinical arena to guide therapy for gonadotropin deficit in males.

1. INTRODUCTION

Reproductive function in humans is governed by the hypothalamic-pituitary-gonadal axis. Among the main players of this axis are the pituitary gonadotropins, follicle-stimulating hormone (FSH), and luteinizing hormone (LH), which together with chorionic gonadotropin (CG) synthesized by the placenta, and thyroid-stimulating hormone produced in the thyrotropes, conform the family of glycoprotein hormones (GPH). All GPHs are heterodimeric proteins that consist of a common α-subunit noncovalently associated with a β-subunit, which is structurally and functionally unique to each member of the GPH family.[1] Both subunits exhibit N-linked (and O-linked as in the case of CG) oligosaccharide chains, whose number will vary depending on the particular GPH and that play a pivotal role in determining several functions of the protein, including circulatory half-life ($t_{1/2}$), in vivo bioactivity, and efficacy to activate their cognate receptor at the target cell.[2–5] The production and secretion of the pituitary gonadotropins are regulated by the action of several hypothalamic, pituitary, and gonadal hormones that modulate the transcriptional activation of their subunit genes[6,7] (Fig. 1), translation of corresponding mRNAs, and secretion of the mature gonadotropins.[4,6]

In the male gonad, the Sertoli cells lining the seminiferous tubules are the primary targets of FSH, while the Leydig cells of the testes are the

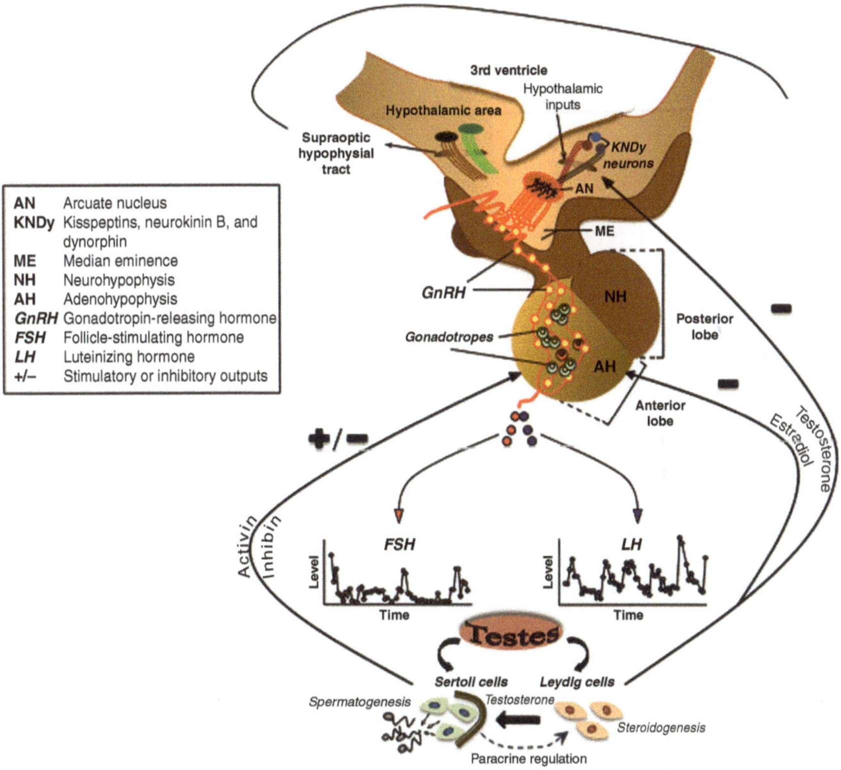

Figure 1 Functional relations of the hypothalamus–pituitary-testicular axis. Gonadotropin-releasing hormone (GnRH) is synthesized and secreted by specialized neurons located mainly in the arcuate nucleus of the medial basal hypothalamus and the preoptic area of the anterior hypothalamus. The synthesis and secretion of GnRH is under the regulation of KNDy neurons. GnRH-producing neurons project to the median eminence where they terminate in an extensive plexus of boutons on the primary portal vessel that delivers GnRH to its target cell, the gonadotrope of the adenohypophysis. The secretion and interaction of GnRH with its cognate receptor occurs in a pulsatile and intermittent manner; such episodic signaling allows distinct rates and patterns of synthesis and pulsatile release of luteinizing hormone and follicle-stimulating hormone. These gonadotropin hormones bind their target receptor in the Leydig cells of the testicular interstitium and the Sertoli cells inside the seminiferous tubules. The transcriptional regulation of the gonadotropin subunit genes as well as the secretion of the heterodimers are controlled by GnRH as well as by the testicular hormones produced in response to gonadotropins. *Modified from Trends Endocrinol Metab. 2010; 21:190–197, and reproduced with permission from Elsevier Ltd.*

primary targets of LH. In these cells, gonadotropins play a key role in regulating testicular function and thereby in male reproduction.[8,9] Receptors for LH and FSH (LH/CGR and FSHR, respectively) have also been detected in a number of extra-gonadal tissues[10,11]; however, the extra-gonadal role of LH and FSH still remains to be more firmly clarified, particularly in humans. In contrast, characterization of multiple CG isoforms, native CG, hyper-glycosylated CG, and the β-subunit of the hyper-glycosylated CG has allowed to assign extra-gonadal/extra-placental functional roles to this gonadotropin in both normal and pathologic conditions[12–15].

Gonadotropins bind their cognate receptors (the LH/CGR and the FSHR) in the testes to stimulate a diverse array of biological effects. In the testes, it is accepted that FSH regulates Sertoli cell number, maturation, and function and thereby (synergically with testosterone) optimal germ cell development.[16–18] Through the action of FSH, Sertoli cells provide physical and metabolic support for germ cell differentiation, meiosis, and transformation to spermatozoa.[16] Nevertheless, the role of FSH as a *sine qua non* in spermatogenesis in humans is still controversial because men with inactivating FSHR mutations have some level of spermatogenesis and may even be fertile.[19] This and other observations[20,21] strongly suggest that FSH and testosterone are both indispensable in man to promote maximal normal sperm production. In the interstitial compartment of the testes, LH is involved in Leydig cell maturation, survival, and proliferation as well as in stimulating androgen production (mainly testosterone) by promoting cholesterol mobilization, StAR (steroidogenic acute regulatory protein, which facilitates the transfer of cholesterol from the outer to the inner mitochondrial membrane) and steroidogenic enzyme expression and activity.[22–24] Under gonadotropin stimulation, the gonads also produce nonsteroid (polypeptide) factors, such as inhibins and activins, which together with the steroid hormones synthesized in response to gonadotropins, regulate the function of the hypothalamic-gonadotrope unit.[25] CG has multiple actions in pregnancy, including binding to the LH/CGR to stimulate fetal Leydig cells and promote testosterone secretion and male phenotypic differentiation.

A number of diseases are associated with gonadotropin deficiency in humans. In some, the cause resides at the hypothalamic level, which is the site of gonadotropin-releasing hormone (GnRH) synthesis (Fig. 1), whereas others originate from the inability of the anterior pituitary to synthesize and/or release one or the two gonadotropins due either to resistance to GnRH, presence of genetic defects, or infiltrative/tumoral disorders of the pituitary gland.

This chapter concentrates on the structure and function of the gonadotropin hormones, mainly LH and FSH, and their application in the clinical arena to treat disease in men.

2. STRUCTURAL FEATURES OF THE GONADOTROPINS

2.1 Gene Structure

The gene encoding the common α-subunit of several mammalian species consists of four exons separated by three intervening introns.[26,27] In humans, this gene is located on chromosome 6q12-q21; the α-subunit mRNA encodes for a protein composed of 116 amino acid residues corresponding to the leader sequence (24 amino acids) and the mature protein (92 amino acids) (Fig. 2).[28] The common α-subunit has the highest sequence homology among mammals, with 76% homology between human and equine and 100% between ovine and bovine sequences.[4,28] The single-copy FSHβ gene subunit is located on chromosome 11p13, consists of three exons and two introns, and codes for a signal peptide of 19 amino acids and the mature protein of 111 amino acid residues. The *FSHB* differs from the other GPH β-subunit genes in that it possesses an extremely long untranslated 3' terminal encoded by exon 3.[29] FSH β-subunit homology among mammalian species is relatively high, with homology values between 92% and 99% and an overall identity of ~77%. In contrast, the multicopy LHβ/CGβ genes are located tandemly in the same chromosome (19q13.32); the *LHB/CGB* cluster is a region that contains seven genes evolved by duplication events of the LHβ gene.[30,31] *LHB* codes a mature protein of 121 amino acids, while the mature CGβ subunit (composed of 145 amino acids) is encoded by at least six nonallelic genes (*CGB*) named *CGB1-3*, *CGB5*, and *CGB7-8*,[30] with *CGB5* and *CGB8* being the most actively transcribed in the placenta, followed by *CGB3*.[31,32] The major site of LHβ and FSHβ subunit expression are the gonadotropes, whereas in the case of CGβ are the syncytiotrophoblastic cells of the placenta. Although both the α- and the β-subunit bind the receptor, receptor recognition and binding specificity resides in the β-subunit.[33]

2.2 Subunit Structure

The tridimensional structures for FSH and CG have been determined.[34,35] Both structures share similar folds in their common α- and β-subunits, which exhibit a cysteine knot motif at their central core of extended hairpin loops. The common α-subunit exhibits five disulfide bridges, while the β-subunit bears six disulfide bridges (Fig. 2). The association between the α- and the β-subunit is stabilized by a disulfide-tethered segment from

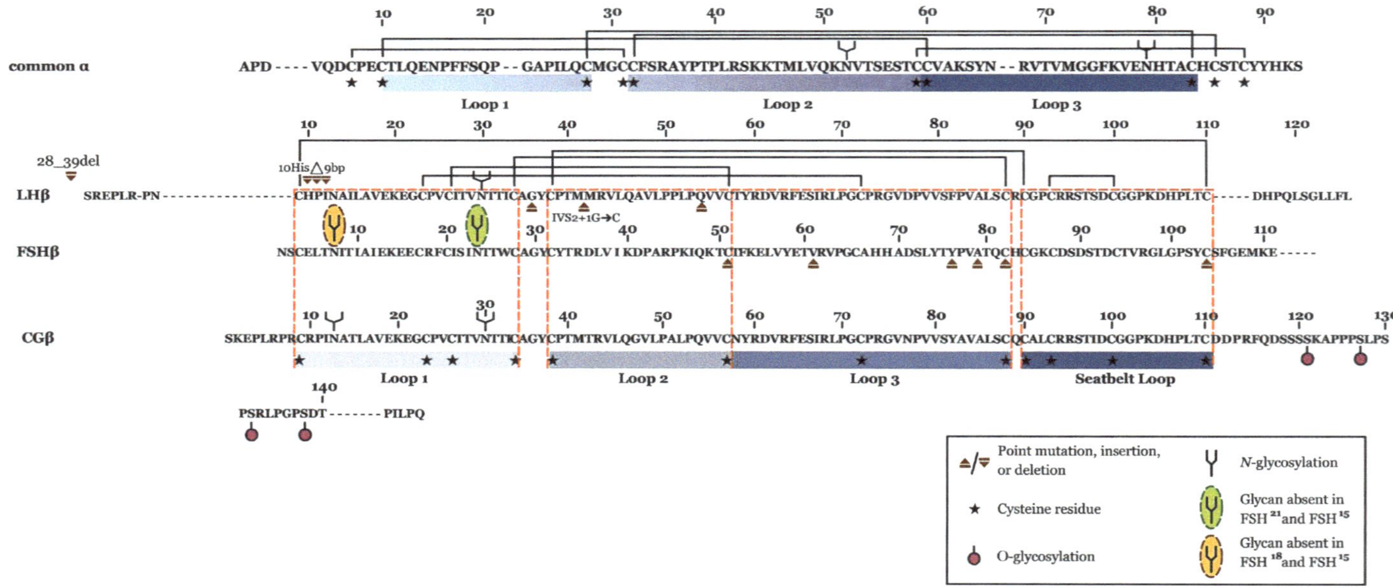

Figure 2 Primary structures of human gonadotropins. The mature amino acid sequences for the human gonadotropin subunits are shown for the primary structures translated from nucleotide sequences. The amino acid sequences were obtained from the National Center for Biotechnology Information (NCBI, USA). Sequence alignments were made using COBALT (NCBI). The location of N-glycans are indicated by the tuning forks above the glycosylation sites. Both FSHβ and CGβ possess tow N-glycans, whereas human LHβ is glycosylated at Asn30. Human CGβ is also O-glycosylated at 4 serine residues 121, 127, 132, and 138). The three loops created by the cysteine knot motifs and the β-subunit seat belt loop are indicated by the bottom blue bars and delineated by perpendicular broken red lines. Cysteine residues are highlighted by stars under subunit sequence. Also shown are the locations of disulfide bonds (brackets at the top of the sequences) and the site of mutations in FSHβ and LHβ.

the β-subunit (the so-called "seat belt") which "embraces" the α-subunit. Some structural differences between FSH and CG exist,[34,35] which are important for discrimination between these two gonadotropins by their cognate receptors. Comparison of the structures of deglycosylated and glycosylated CG and glycosylated FSH, indicates that there is no influence of the oligosaccharides on the backbone conformation of the proteins, although the structures predict contacts between the glycan attached to αAsn52 (see later) and neighboring residues at the β-subunit (e.g., Tyr58 in FSHβ), suggesting that this particular glycan may contribute to dimer stability.[36]

As mentioned earlier, the α- and β-subunits of the gonadotropins are decorated by N-linked (and O-linked in CGβ) oligosaccharides that play a pivotal role in determining several functions of the protein[4,5,37] (Fig. 2). In fact, glycosylation is the most important posttranslational modification in GPHs since it influences many processes. Some of the known functions of the GPHs include subunit folding, heterodimerization, stability, sorting, transport, and conformational maturation of the dimeric proteins within the cell, as well as the efficiency of heterodimer secretion. Glycosylation is essential in determining the metabolic fate of the hormone and may also influence the interaction with its cognate receptor and selective activation of particular signaling pathways[38–49].

The common α-subunit exhibits two N-linked glycans, at positions Asn52 and Asn78. FSH β-subunit bears also two N-linked oligosaccharide chains at positions Asn7 and Asn24, while LHβ has only one glycan attached to residue Asn30. In CGβ two N-linked carbohydrate chains are located at positions Asn13 and Asn30 and four O-linked glycosylation sites at serines 121, 127, 132, and 138 (Fig. 2).[2] Although terminally sulfated glycans are characteristic of LH,[50,51] human LH is an exception, as 22–39% of the glycans terminate with α2-3 and α2-6 linked sialic acid.[52] Human FSH is minimally sulfated (sulfated glycans are less than 10% of total glycan population in this gonadotropin)[53] and glycans terminating in α2-6 and α2-3 linked sialic acid predominate. Since Chinese hamster ovary (CHO) cells lack the GalNac-transferase and sulfotransferase enzymes necessary for terminal sulfate transfer during LH processing,[54] recombinant human LH produced by these cells is mostly sialylated.[55] Differences in the extent of sialylation and sulfation among these gonadotropins are of paramount importance for determining both the in vivo bioavailability and the bioactivity of the hormone; terminal sialic acid prolongs while sulfate residues shorten the plasma $t_{1/2}$ of the glycoprotein in blood.[56]

In general, oligosaccharides on the β-subunits play a major role in determining the metabolic clearance rate of gonadotropins,[49] whereas the glycan in position αAsn52 is more important for the activation of the receptor/signal transducer (G proteins) system and the subsequent biological response.[33,43,44,57–60] A potential role for the αAsn78 glycan in FSH-mediated signal transduction has also been proposed by two mutagenesis studies.[61,62]

As described earlier, the human FSHβ subunit is glycosylated at residues Asn7 and Asn24, and differences in the number of glycans attached on this subunit (none, one, or two) constitutes the basis of the macroheterogeneity of the α/β dimer.[63] In vitro and in silico studies have shown that hypoglycosylated pituitary human FSH preparations exhibit higher FSH receptor binding activity and bioactivity as compared to fully glycosylated FSH.[47,48,63] Thus, similar to the influence of microheterogeneity (determined by the carbohydrate composition of the glycans attached to the protein core) on the ability of the gonadotropins to activate and trigger intracellular signaling,[63,64] it seems that the extent of glycosylation of the FSHβ subunit may also contribute to its bioactivity. It appears that differences in FSH glycosylation influence the stability of binding to the FSHR, with a more stable FSHR interaction of the hypoglycosylated FSH glycoform relative to the fully glycosylated one; these differences in FSH glycoform binding suggest mechanisms for the variant biological effects of fully and partially glycosylated FSH in vivo and in vitro.[47,63,65] Macroheterogeneity of FSH is important from the physiological point of view given that recent evidence has shown that fully glycosylated FSH represents approximately 80% of FSH in pooled pituitary and urinary FSH samples from postmenopausal women, while partially glycosylated FSH represents 52–70% of the samples isolated from pituitaries derived from autopsies of women in reproductive age.[63,66] Further, the abundance of the low molecular weight glycoform, FSH^{21}, is correlated with the age of the donor: this glycoform is more abundant in pituitaries of younger women and decreases over the reproductive lifespan, such that in the postmenopausal age FSH^{24} (fully glycosylated) is the dominant glycoform.[67] Recent studies on circulating gonadotropins also indicate that the levels of hypo- and fully glycosylated FSH and LH vary across the menstrual cycle, suggesting that glycosylation is under hormonal control.[68] Recombinant diglycosylated human FSH was found to be cleared from the circulation faster than wild-type, tetra-glycosylated recombinant FSH[49] and studies in human pituitary extracts reported that pituitary FSH from young, cycling women was cleared more rapidly than FSH from postmenopausal women.[69] Concurrently, these studies indicate that the difference in clearance rates between hypo- and tetra-glycosylated FSH reflects, at least in part, the high abundance of more rapidly cleared hypo-glycosylated FSH in young women and its low abundance in postmenopausal women. These findings may have important implications for the pharmacological treatment of infertility in assisted reproduction.

2.3 Polymorphisms and Mutations in the β-Subunit Genes

Several polymorphisms in the LH and FSH β-subunit genes of the gonadotropins, associated with potential impact on gonadotropin action have been described. These include the Trp8Arg/Ile15Thr, Gly102Ser, and Ala-3Thr polymorphisms in the *LHB*, and the -211G/T polymorphism in *FSHB*. The most widely studied *LHB* polymorphism is the Trp8Arg/Ile15Thr combination in exon 2, which is in complete linkage disequilibrium.[70,71] This polymorphism is world-wide distributed and presents varying frequencies depending on the particular ethnic group screened. The polymorphic LH molecule (named LH-V) is not recognized by specific immunoassays (particularly fluoroimmunometric assays) nor is, in general, strongly linked with fertility or endocrine disturbances, albeit some few studies suggest that it might do so.[70,72–75] The glycosylation of this LH variant differs from that exhibited by the wild-type variant in the introduction of an additional glycosylation site at Asn13β, which leads to differences in plasma-half life and in vitro biological activity.[75] The Gly102Ser in exon 3 of *LHB* is a low frequency polymorphism that has been associated in some populations with infertility in men and menstrual disturbances in women.[76–78] Finally, in the Ala-3Thr variant, three residues are changed before the signal peptide cleavage[79]; this is an interesting variant that has been reported to behave as a biased agonist in vitro, being less potent to stimulate cAMP production and more potent to provoke phosphatidylinositol turnover.

The -211G/T *FSHB* variant in the promoter region of the subunit gene has been associated with varying levels of circulating FSH; in several populations, TT-genotype carriers exhibit lower serum FSH, inhibin B, and testosterone levels as well as lower testicular volume than the GG carriers, and the former genotype has been associated with infertility in men.[80]

Structural alterations in gonadotropins due to mutations in β-subunits leading to gonadotropin deficiency are very rare. When they occur in LHβ, they are clinically manifested by delayed puberty and infertility in homozygous males and infertility in affected women.[81–87] The fact that males bearing homozygous mutations in *LHB* exhibit a normal male phenotype, illustrates the role of CG during fetal life on Leydig cell proliferation, differentiation, and androgen production. A number of mutations in LHβ have been described. The missense mutation Gly36Asp, disrupts a five amino acid motif (CAGYC) important for the formation of the cysteine knot.[84,85] Thus, the abnormal subunit does not heterodimerize with the common α-subunit. The male subject bearing this mutation presented hypogonadism, delayed puberty, undetectable serum LH, low testosterone levels, and azoospermia, despite elevated serum FSH, underlying the critical role of intratesticular testosterone for normal spermatogenesis. In the Gln54Arg mutation, heterodimerization of α/β subunits occurs, however, the resulting dimer does not bind to its cognate receptor.[86] The reported adult homozygous male patient with this mutation exhibited absence of puberty and infertility, with high concentrations of immunoactive LH, while in heterozygous males puberty was normal but fertility reduced. Heterozygous women with this mutation exhibited no reproductive abnormalities.[86] The IVS2+1G→C, which disrupts intron 2 splicing, leads to inclusion of the entire intron and disruption of exon 3 reading frame, yielding a highly abnormal mRNA and, theoretically, an aberrant protein with additional 79 amino acids inserted after the Met41 residue, a sequence unable to form the seatbelt loop of the subunit.[83] Male subjects bearing this mutation presented delayed puberty and infertility, whereas the affected woman presented normal puberty but secondary amenorrhea and infertility; serum LH was undetectable and testosterone levels were very low. The twelve nucleotide deletion (28_39del) at the LHβ signal peptide leads to deletion of four leucines at positions 10–13 (Leu10Δ12 bp), and thus in failure for normal β-subunit processing.[82] This mutation was reported in a compound heterozygous man who also had the IVS2+1G→C mutation described earlier. The two sisters of the proband had the same mutations and presented normal puberty and oligomenorrhea.[82] Finally, the nine-base deletion in exon 2, which impairs cysteine knot formation, leads to a β-subunit lacking amino acids 10–12 and a LH molecule with residual activity sufficient to allow reaching adequate intratesticular concentrations of testosterone. Thus, the affected man with this mutation exhibited normal spermatogenesis.[81]

FSHB mutations are also rare; male homozygous subjects bearing mutations in FSHβ present normal virilization during puberty and azoospermia,[88–90] while homozygous females exhibit absent or partial puberty and infertility.[88,91–95] The phenotype in men with mutations in *FSHB* contrasts with its counterpart in mice, in which absence of this subunit (or the FSHR) does not lead to infertility.[96,97] It also contrasts with findings in men bearing loss-of-function mutations in the *FSHR* gene, in whom fertility is preserved.[19] Mutations in *FSHB* occur in exon 3 and lead to impaired disulfide bonds formation, severely altered tridimensional structure, and inability of the FSH β-subunit to heterodimerize with the common α-subunit. The *FSHB* mutations described so far include the Val61X mutation, in which a two-nucleotide deletion in codon 61 affects the amino acid sequence 61–86 followed by a premature stop codon (Val61Δ2 bp/87X), eliminating amino acids 87–111 of the subunit[90,95,98]; the Cys51Gly mutation, which was reported in a compound heterozygous women bearing also the Val61X mutation[98]; the Tyr76X nonsense mutation, which predicts a translated protein missing amino acids 76–111[88]; the Ala79X, a one nucleotide deletion at codon 79 (c.289delG) leading to a frameshift and a premature stop at codon 109 (A79fs108X)[92]; the Cys82Arg mutation, which similar to the Cys51Gly substitution alters the cysteine knot structure of the β-subunit[89]; and the Cys122Arg (Cys104Arg without considering the signal peptide) mutation,[99] which disrupts a disulfide bond that is part of the "seat belt" that wraps around the α-subunit.[34]

No germline loss-of-function mutations have been reported in the common-α subunit gene, which supports the concept on the evolutionary conservation of this subunit as well as its importance for pregnancy and embryo survival, given that mutations in this or the CGβ subunit would yield inactive CG.

3. REGULATION OF THE SYNTHESIS AND SECRETION OF THE PITUITARY GONADOTROPINS

The control of LH and FSH synthesis and secretion is complex and involves a coordinated interplay between the gonads, the pituitary, and the hypothalamus. The hypothalamic component is represented by GnRH, a decapeptide synthesized by specialized neurons predominantly located in the arcuate nucleus of the mediobasal hypothalamus and in the preoptic area of the anterior hypothalamus. The axons of these neurons are projected either to various regions of the brain, where GnRH acts as a neurotransmitter or neuromodulator of reproductive behavior or to the median eminence, where GnRH enters the portal circulation to reach the pituitary gonadotropes

(Fig. 1).[6] The release of GnRH into the hypophyseal portal circulation occurs in a pulsatile manner[100–102]; the episodic GnRH secretion is finely regulated by the coordinated interaction between hypothalamic neurons of the arcuate nucleus that coexpress kisspeptins, neurokinin B and dynorphin (KNDy neurons), and the GnRH-producing neurons (Fig. 1). Neurokinin-B and dynorphin operate reciprocally as major positive and negative regulators of the episodic release of kisspeptin, which in turn is the effector of this neuronal complex for the stimulation of the neurosecretory activity of GnRH neurons via interaction with its cognate receptor, KISS1R.[103] Once in the gonadotroph cells, GnRH interacts with its cognate receptor (GnRHR) to induce gene expression as well as stimulus-secretion of gonadotropins.[104] In fact, GnRH-regulated expression of gonadotropin α- and β-subunits provides a genomic signature unique to functional gonadotropes.

Many pathways need to coalesce to achieve the complexity and, more importantly, the differential regulation of LH and FSH subunits expression, a first step to the production and secretion of the gonadotropins. The activation of signaling pathways promoting the interaction of distinct transcription factors with specific response elements present in the gonadotropin subunit promoters to control differential gonadotropin gene transcription vary depending on the cell context and experimental conditions.[105–108] In primary pituitary cultures, G-GH3, and LβT2 cells, the GnRHR couples to G_s and $G_{q/11}$ but not G_i, whereas in αT3-1 pituitary precursor cells, CHO-K1 cells, and COS-7 cells, the GnRHR appears to couple almost exclusively to $G_{q/11}$.[105,109–111] Coupling to other G proteins has also been reported.[112,113] Consequently, increases in a number of distinct second messengers including cAMP, inositol 1,4,5-trisphosphate, Ca^{2+}, and diacylglycerol, as well as activation of multiple protein kinase-mediated signaling pathways following GnRH exposure have been shown to be important for decoding the GnRH signal by the gonadotrope and evoke frequency-dependent, differential LH and FSH synthesis and secretion.[105,106,111,113–116] Some of these studies, however, have used acute tonic treatment rather than pulsatile stimulation, which is known to be fundamental for gonadotropin gene transcription.[117,118] More specifically, the gonadotrope responds to pulsatile GnRH stimulation activating signaling cascades mediated by MAPKs (mitogen-activated protein kinases), including ERK1/2, (extracellular signal related kinase), JNK (jun N-terminal kinase), and p38.[119,120] In particular, it seems that ERK1/2 phosphorylation and translocation to the nucleus is involved in differential transcriptional regulation of LHβ and FSHβ genes via GnRH-provoked activation of MAPK phosphatases, particularly

DUSP1 (dual-specificity kinase), which allow for differential activation patterns of these kinases.[121,122] Other signaling pathways involved in differential synthesis of LHβ and FSHβ subunits include those dependent of the PKA/cAMP pathway (see later), the Ca^{2+}/calmodulin/calcineurin pathway, which allows nuclear translocation of the nuclear factor of activated T-cells (NFAT) upon cell exposure to GnRH,[123–126] and the WNT/β-catenin signaling pathway.[127]

3.1 Transcriptional Regulation of the Common α-Subunit Gene (αGSU), LHβ (*LHB*), and FSHβ (*FSHB*) Genes

Although the α-subunit is produced in excess of the gonadotropin β-subunits and αGSU transcription does not appear to be tightly regulated by GnRH pulse frequency as occurs with the β-subunits, some studies have indicated that α-subunit transcription is favored by fast-frequency (interpulse interval 8–30 min) GnRH pulses[128,129] and that it can be regulated by gonadal factors.[130] In the human gonadotrope, the α-subunit promoter exhibits two tandemly repeated cAMP/PKA response elements (CRE) [at positions −146 to −111 relative to the transcription initiation site][131] that bind the nuclear CRE-binding proteins, including CREB, CRE modulator, and c-Jun/ATF2 heterodimers to activate gene transcription. Other regulatory elements involved in basal transcription of the αGSU gene include the homeobox 1 (Pitx1) response element[132,133] and GSE (gonadotrope-specific element),[134–136] as well as the GnRH responsive unit at the distal 5' regulatory region (which also define αGSU gene expression in a cell-specific manner) comprised by several elements including PGBE, αBE1 and 2, and Ets binding sequence; these elements bind several factors including members of the LIM homeodomain family αBP1 and 2, and Ets.[133,137] Meanwhile, the αGSU promoter regions implicated in GnRH-regulated αGSU transcription mainly include PGBE, αBE, and the Ets sequence, with CRE at the proximal promoter region interacting with the former elements.

The LHβ subunit gene transcription is stimulated by high frequency GnRH pulses [interpulse interval ~60 min in in vitro systems],[138] which in turn provoke MAPK activation and LHβ transcription through rapid induction of the immediate early gene Egr-1, which interacts with basal factors to induce LHβ transcription.[133] In rodents, Egr-1 sites are located in tandem with SF-1 sites at either site of homeodomain element (HD) that interacts with Pxt1; these three elements, which act synergistically to promote GnRH-induced LHβ transcription also interact with β-catenin to increase transcription of this subunit.[127] Meanwhile, the scaffold protein p300 allows interaction between these proximal promoter sites and the more distally located SP-1 and CArG elements, also involved in pulsatile LHβ transcription.[139] Apparently, the GnRH frequency-regulated LHβ transcription is governed by the interplay between ubiquitin, which establishes on-off cycling of SF-1 and Erg-1 factors at the LHβ promoter, and the Egr repressor proteins Nab1/2 (NGFI-A binding proteins) which are induced by low frequency GnRH pulses.[138]

Several transcription factors act in concert to activate cell specific and hormone-regulated differential expression of the gonadotropin β-subunits. In FSHβ, synthesis of this subunit represents the rate-limiting step for FSH dimer synthesis and constitutive FSH secretion from the gonadotrope. The FSHβ gene promoter contains in its 5' proximal flanking region a half cAMP-response element/AP1 site, which binds CREB as well as c-Fos and c-Jun, both members of the AP1 family. In fact, the CREB consensus binding sequence (TGACGTCA) is very similar to that of AP1. This GnRH-responsive element and inducible cAMP early repressor (ICER) have been implicated in the regulation of murine FSHβ gene expression. CREB can bind to cAMP response elements on the FSHβ gene promoter and could give rise to the differential GnRH pulse effect on gonadotropin gene expression by recruiting CREB binding protein (CBP) at slow pulse frequency of the GnRH decapeptide. In this scenario, high pulse frequency of GnRH may increase ICER which binds to CRE on the FSHβ gene promoter thereby dampening gene transcription.[104] Thus the induction of this and other negative regulatory transcription factors (such as SKIL and TGIF1)[140] could potentially provide a mechanism for the differential regulation of gonadotropin gene expression under low and high GnRH pulse frequency, with low pulse (interpulse interval 120–240 min) favoring and high pulse repressing FSHβ gene transcription. Distal to the GnRH-responsive element described above is a *cis*-element that binds Nur77, which seems to regulate GnRH-induced derepression of the murine FSHβ gene.[141,142]

The dynamics of gonadotropin synthesis is complex and includes the action of an additional number of factors (endocrine and paracrine/autocrine) that may directly influence on the gonadotrope response to GnRH. It is recognized that sex steroids modulate gonadotropin synthesis and secretion at the hypothalamic and the pituitary levels. Estrogens exert both positive and negative feedback on gonadotropin synthesis and secretion; the negative feedback occurs mainly at the hypothalamic level through influencing KNDy neurons and thereby GnRH secretion.[143,144] The positive feedback of estrogens also occur through actions at the hypothalamus[144] and it has also been shown that estrogens acting via the estrogen receptor-α, promote LHβ gene transcription either directly through binding an imperfect estrogen response element in the distal LHβ promoter[145] or interacting with SF-1 and Pitx1,[146] or indirectly by influencing expression of activators [e.g., Egr-1) and repressors of transcription.[147] Estrogens exert a suppressive effect on FSHβ gene transcription by interaction with the estrogen receptor-α but the mechanisms through which this is effected are still unclear. Progesterone, another steroid hormone produced by the gonads, mainly by the corpus luteum, and the placenta regulates gonadotropin synthesis and secretion acting at both the hypothalamus and the pituitary. The effects at the gonadotrope level have been difficult to analyze due to the induction of the progesterone receptor by estrogen administration. Nonetheless, in LβT2 cells overexpressing the progesterone receptor, exposure to this steroid hormone suppressed basal and GnRH-induced LHβ gene expression, an effect that required the presence of elements within −300/−150 of the LHβ promoter but not a direct interaction of the steroid-progesterone receptor B complex with the promoter.[148,149] By contrast, progesterone exerts facilitatory effects on FSHβ gene transcription through direct interactions of the activated progesterone receptor with hormone response elements of the FSHβ promoter.[150] A similar effect has been observed for androgens and glucocorticoids, which down regulate LHβ and up regulate FSHβ transcriptional activation through direct interactions of the activated androgen receptor with either rat Sp1 or bovine SF-1 factors (to repress GnRH-induced LHβ transcription) or with hormone response elements (to promote FSHβ gene transcription).[133,150,151] In contrast to estrogens, androgens suppress transcriptional activation of the αGSU gene through interaction with CRE-binding proteins in both human and mouse αGSU promoters.[133]

Nonsteroid factors also regulate gonadotropin expression. These include: (1) leptin,[152–154] which increases the synthesis and secretion of LH and FSH through its effects on the transcriptional regulation of LHβ and FSHβ genes mediated by the LIM-homeodomain transcription factor Isl-1 gene[155]; (2) Activin, growth differentiation factor 9 (GDF9), and bone morphogenetic proteins (BMP, specifically BMP4, 6, and 7), all members of the transforming growth factor β (TGFβ) superfamily, which act synergically with GnRH to regulate the transcriptional activation and mRNA expression of the FSHβ gene in an autocrine manner[25,156–164]; (3) Follistatin, which binds activin thereby inhibiting its effects on the gonadotrope[165]; (4) Inhibin, which prevents activation of the activin-mediated signaling pathway and consequently FSHβ gene transcription[166,167]; and (5) Pituitary adenylate cyclase-activating polypeptide, a member of the vasoactive intestinal peptide-secretin-glucagon peptide superfamily, which either alone or with GnRH stimulates expression of each gonadotropin subunit as well as transcription of the GnRHR gene.[168–170]

4. GONADOTROPIN SECRETION

As discussed in the preceding sections, the differential regulation of LH and FSH synthesis and secretion is modulated by GnRH pulse frequency and by the concomitant effects of sex steroids and nonsteroid factors. In contrast to LH, which is stored in complex secretory granules, secreted in pulses, and associated with a midcycle release from the gonadotrope upon GnRH stimulation, FSH is transported to the plasma membrane in vesicles and then released to the circulation. Thus, the secretion of FSH is constitutive, whereas that of LH is regulated by GnRH. In vitro studies have determined that this difference is due to a hydrophobic stretch of amino acids (LSGLLFL) in the LHβ subunit tail (Fig. 2).[171] Further, FSH secretion could be switched from the constitutive to the regulated pathway by fusing that sequence to the carboxyl-terminal subunit tail of FSHβ.[172,173] Nevertheless, several studies have found that although the constitutive (basal) component of FSH secretion predominates, low-frequency GnRH stimulation leads to coincident release of FSH in discrete pulses, which are difficult to detect in peripheral circulation given the low concentration of FSH and its relatively long plasma $t_{1/2}$, unless deconvolution techniques are employed to identify true FSH pulses.[174–176] This pulsatile release of FSH correlates with the transcriptional effects of GnRH on *FSHB* described earlier.

In the majority of mammalian species studied, the GnRH-dependent secretion of LH occurs in a pulsatile manner, and the frequency of pulses detected in peripheral circulation changes depending on the phase of the menstrual cycle.[177–179] In regular cycling women, the pulse frequency of LH varies from every 90–100 min during the early follicular phase to one pulse every 4–8 h during the late luteal phase.[177,180,181] Variations in LH pulse frequency are mainly regulated by the changing levels of estradiol and progesterone that occur during follicular maturation and the functional life span of the corpus luteum.

In men, pulsatile LH release can be detected early in infancy, particularly during the first 3–6 months after birth (i.e., during the minipuberty of the neonate), but also in childhood; thereafter, significant changes in LH secretion occur until 2 years before puberty, where sleep entrained LH secretion becomes apparent. From mid-childhood to adulthood the production rate of LH progressively increases and it is accompanied by a modest increment in pulse frequency.[182] In men, GnRH differentially controls the release of FSH and LH in fetal life, infancy, puberty, and adulthood. In fact, the frequency and amplitude of LH pulses increase from prepuberty to puberty, while that of FSH remains constant.[183] Studies in normal and hypogonadal prepubertal boys have also shown that the testes are involved in the regulation of FSH but not LH levels during prepuberty.[184] In adult men, LH pulses occur approximately every 55 min.[185,186] The synthesis and secretion of FSH are regulated by several factors including activin A, inhibin B, and estradiol concentrations,[25,160,187,188] whereas testosterone is the main driving force that regulates LH production, either directly or via aromatization to estradiol.[188,189] In aging men, the levels of testicular androgens decline, and several functional abnormalities in the reproductive axis can be detected, including impaired testicular response to LH, attenuated pituitary bioactive LH reserve, release of high-frequency/low amplitude LH pulses, and disorderly LH release due to altered GnRH output and/or defective negative feedback by testosterone.[190–194]

5. PHARMACOKINETICS OF GONADOTROPINS

As discussed earlier, carbohydrates in GPHs play a pivotal role on the metabolic clearance rate and the in vivo biological potency of the gonadotropins. These functions are highly dependent on the amount of glycans attached to the protein and the specific type of terminal residues present in their oligosaccharide structures. Glycans in FSH and CG predominantly terminate in sialic acid; this sugar, and particularly the number of exposed terminal galactose residues are key factors in determining the hormone's survival in the circulation.[195] Exposure of terminal galactose residues increases glycoprotein clearance from plasma by a mechanism that involves hepatocyte receptors for the galactose-terminated complex molecules. However, this mechanism is irrelevant for CG clearance, as blocking the hepatocyte receptors with asialoglycoproteins has no effect on intact CG clearance[196] and kidney clearance is apparently a more important mechanism.[197] Highly sialylated glycoproteins circulate for longer periods than those bearing less sialylated or sulfated glycans,[6,198–201] whereas sulfated glycans, such as those present in LH and minimally in FSH, or oligosaccharides bearing terminal mannose or N-acetylglucosamine residues, accelerate clearance of the molecule by specific receptors expressed by hepatic endothelial and Kupffer cells.[202,203] The plasma $t_{1/2}$ of FSH and CG is therefore longer than that of LH. Recombinant asialo-gonadotropin variants as well as deglycosylated FSH are rapidly cleared from the circulation and thus are practically inactive in vivo when compared to the intact variants.[204] Conversely, recombinant FSH (recFSH) containing a hybrid β-subunit composed of the β-subunit of FSH and the carboxyl-terminal peptide of CGβ [(FSH-CTP), which is rich in O-linked sialylated glycans (Fig. 2)], or to which additional glycosylation sites have been added, exhibit a prolonged plasma $t_{1/2}$ (approximately 2–4 times longer) than intact recombinant FSH.[205–208]

The plasma disappearance of endogenous LH is slower in postmenopausal than in young women,[209] which is likely due to the increased sialylation of gonadotropins that occurs during the postmenopause.[210–212] Gonadotropins are also degraded in the kidney (as demonstrated for CG) albeit the specific structural requirements that promote their renal uptake are still unknown. Gonadotropin molecules that have not been metabolized by the liver are excreted in urine in forms that still are highly bioactive. This characteristic has allowed the isolation, purification, and clinical application of urinary gonadotropins from pregnant and postmenopausal women for inducing virilization and spermatogenesis as well as for infertility treatment (see later). The free subunits of gonadotropins are more rapidly eliminated from the circulation than the desialylated dimers or molecules with fused subunits, which suggests that pathways different to those described earlier also determine the metabolic fate of gonadotropins.

A number of studies on gonadotropins have been performed to determine their pharmacokinetic behavior. Nevertheless, the results may vary depending on the specific preparation studied, methods of measurement of the circulating hormone, parameters reported, and routes of administration. In general, the plasma $t_{1/2}$ of CG is longer than that exhibited by LH, whereas that of FSH is intermediate between CG and LH. The pharmacokinetic behavior of gonadotropins has been extensively studied.[213–221] Analyzed as a single exponential, the initial plasma $t_{1/2}$ of urinary FSH after its intravenous (i.v.) administration varies from 4.7 to 6.9 h, whereas for biexponential kinetics the initial $t_{1/2}$ ranges from 1.5 to 2 h and the slow phase $t_{1/2}$ from 15 to 17 h. Despite differences in sialylation between urinary and recombinant FSH produced by CHO cells,[222] the pharmacokinetics of both FSH preparations is quite similar, indicating that such differences in terminal sialylation does not significantly impact on the metabolic clearance of the hormone injected, at least in the form of administration employed in these studies.[214] Urinary and recombinant FSH are eliminated in the urine in proportions of 20% and 10% of the initial dose administered, respectively, and their estimated total clearance is approximately 0.5 L/h.[216] These values, which are below the glomerular filtration rate, suggest either that FSH could be reabsorbed after filtration, that the glycosylated protein is too large to be freely excreted or that the FSH molecule could be subjected to a certain degree of metabolism by the kidney. The subcutaneous (s.c.) administration of highly purified preparations of urinary or recombinant human FSH produced by CHO cells yielded nearly similar elimination half-lives (around 30–40 h) and bioavailability values.[216,219–221]

The pharmacokinetic parameters of recombinant (CHO cells-produced) and urinary LH from postmenopausal women are quite similar, which is in agreement with the higher sialic acid content of the former.[55] Following i.v. administration, the initial and terminal plasma $t_{1/2}$ of both preparations approximate 1.2 h and 10–12 h, respectively. The total plasma clearance rate of urinary LH is ~1.2 L/h, while that of recombinant preparation is ~1.7 L/h; renal clearances are 0.3 ± 0.2 L/h and 0.03 + 0.02 L/h, respectively.[218] The contribution of the kidney to the total clearance of urinary LH is ~30%, a value which is significantly higher than that for recombinant LH (~4%); this difference may probably be due to the presence of CG in the urinary LH preparation (see later). Despite these differences, the bioavailability of both preparations is quite similar.[217]

The net biological potency of different gonadotropin preparations in in vivo conditions is intimately related with their circulatory $t_{1/2}$, albeit other factors such as glycan composition and receptor binding capacity of their corresponding variants seem also important.[198,222,223] In fact, despite that recombinant and urinary-derived human FSH preparations exhibit similar pharmacokinetic features and clinical profile in assisted reproduction therapy,[224] administration of recombinant FSH to hypogonadal men led to higher biological-to-immunological FSH ratios than administration of urinary FSH, emphasizing on the role of FSH glycans in determining the intrinsic bioactivity of the gonadotropin.[223]

The kinetics of endogenous CG has been mainly determined by studying its clearance from the circulation immediately after term placental extraction. These studies have shown that CG circulating in term pregnancies has a short initial plasma $t_{1/2}$ of approximately 5 h and a terminal $t_{1/2}$ between 30 and 36 h. In women that underwent legal abortion during the first trimester of pregnancy, CG showed a longer plasma $t_{1/2}$, 30 and 65 h, probably due to the presence of highly sialylated isoforms during the early stages of pregnancy.[199] The plasma $t_{1/2}$ of CG after its intramuscular administration approximates 60 h. The extended plasma $t_{1/2}$ of CG makes this gonadotropin the preferred preparation when prolonged administration of LH and long-term effects on androgen production are needed for therapeutic purposes as will be discussed later.

6. GONADOTROPIN FORMULATIONS AVAILABLE FOR TREATING HYPOGONADOTROPIC HYPOGONADISM AND INFERTILITY IN MEN

Adolescents and adult male patients with congenital hypogonadotropic hypogonadism (HH) (Table 1) should be initially treated with the combination of FSH and LH (or CG), to promote Sertoli cell maturation and spermatogenesis as well as development of secondary sex characteristics. As will be discussed in the following sections, other indications for gonadotropin administration as a therapeutic strategy include acquired HH and normogonadotropic oligozoospermia.

In the market, there are currently a number of available preparations that contain LH, FSH, or CG alone or in mixture. hMG (human menopausal

Table 1 Causes of Hypogonadotrophic Hypogonadism (HH).

Congenital forms
1. Isolated GnRH deficiency (normosmic HH and hiposmic/anosmic HH or Kallman syndrome)
2. Gonadotropin-releasing hormone receptor mutations
3. Mutations in *FSHB* or *LHB*
4. Combined pituitary hormone deficiency
5. Genetic syndromes (e.g., Prader-Labhart-Willi, CHARGE, Lawrence-Moon-Bardet-Biedl)
Acquired forms
1. Tumors (pituitary adenoma including prolactinoma, craniopharyngioma, germinomas, meningiomas, gliomas, and astrocytomas)
2. Infiltrative disease (hemochromatosis, granulomatous disease, histiocytosis, and sarcoidosis)
3. Iatrogenic (traumatic pituitary stalk interruption syndrome, irradiation, high dose corticosteroids, and anabolic steroids)

gonadotropin or menotropin), the first commercialized preparation, contains both FSH and LH as well as some quantities of CG from pituitary origin; CG is usually added to this preparation in order to accomplish with the pharmacopeia requirements and to reach the desired level of LH activity (FSH:LH ratio of 1:1). In contrast to the conventional hMG preparations that contain <5% of pure gonadotropins, highly purified preparations (HP-hMG) with higher specific activity (2000–2500 vs. ~100 IU/mg protein) also are currently available, which allows its s.c. administration. Application of immunological-based techniques with highly specific monoclonal antibodies towards FSH, has made possible to obtain an urinary FSH preparation with very low (≤0.1 IU) LH activity and high specific activity (10,000 IU/mg protein) amenable to s.c. administration. Also from urinary origin is CG obtained from urine of pregnant women, which has a safe clinical profile but that, as non-HP hMG, is contaminated with non-CG proteins and thus should be administered intramuscularly. The current gonadotropin marketplace has been enriched with several recombinant preparations. These include recFSH, recLH, recCG, long-acting recFSH (corifollitropin alpha), and a mixture of recLH/recFSH. The first preparation available for treatment in the clinical setting was recFSH. Two similar recFSH preparations, whose production differs in the technical approach employed to express the heterodimer in CHO cells and that exhibit slight differences in posttranslational modifications are available (follitropin alpha and beta). Both formulations are highly pure, possess the highest specific activity among the available FSH products (between 7000 and 13500 IU/mg protein), and are completely devoid of LH activity. Recombinant LH and CG are also available for clinical use with the advantages of bearing high specific activity as well as purity. Recombinant gonadotropins differ from the natural gonadotropins in the glycans attached to the protein core as CHO cells do not express α2-3 sialyltransferase nor the sulfotransferase necessary for terminal sulfation of LH (and FSH to some extent). To overcome this problem, several pharmaceutical companies are currently developing recFSH formulations produced in human cell lines, which exhibit an identical glycosylation pattern that naturally occur in FSH. Preliminary studies have suggested that these preparations may also be eventually added to the therapeutic arsenal for the pharmacological management of infertility in both women and men.[225–227]

Finally, corifollitropin alfa is an interesting formulation that consists of a recombinant fusion protein of recFSH and the carboxyl-terminal peptide of CGβ (Fig. 2), which possesses four O-linked sialic acid-enriched glycosylation sites and thus exhibits a longer plasma $t_{1/2}$ than recFSH [~65 h compared with ~35 h for recFSH].[228–230] Although this formulation has been successfully used in women undergoing controlled ovarian stimulation, the experience with this compound to stimulate spermatogenesis in men is very limited.

Other gonadotropin or related compounds are currently under development. Of particular importance is the development of allosteric modulators of the gonadotropin receptors designed to regulate the intensity of the gonadotropin stimulus and even provoke biased signaling at the receptor level.[33,46,231–235] A detailed description of these compounds, however, is not within the scope of this chapter. With this background information, now we can review the indications for administering gonadotropins as a therapeutic strategy in men.

7. TREATMENT WITH GONADOTROPINS IN THE MALE

The primary indication of gonadotropin administration in males is HH with functional testes. In these patients, the clinical and laboratory features will depend on the specific etiology of the HH, the degree of hypothalamic, gonadotrope, and testicular dysfunction, and the age of the patient at the time of diagnosis.[236] The causes of HH amenable to gonadotropin therapy are summarized in Table 1. Other indication for gonadotropin therapy is normogonadotropic oligospermia, which often leads to infertility.

The hypogonadal state may be identified neonatally, during childhood, at pubertal age or in adulthood. Given that exposure to gonadotropins during fetal life or neonatally has long lasting effects on testicular function that may become apparent until adulthood, it has been suggested that treatment with gonadotropins early after the diagnosis of HH may significantly improve the fertility potential of the patients.[237–240] In fact, treatment with gonadotropins in azoospermic men with postnatal or postpubertal acquired HH is more effective in terms of pregnancy rate than in men with congenital HH, indicating that exposure to gonadotropins during fetal life, the neonatal period (i.e., the so-called "minipuberty" of the neonate) and in puberty is a key factor for reaching normal spermatogenesis in adulthood.[240–242]

7.1 Gonadotropin Administration in Neonatal Patients with HH

The absence of pituitary gonadotropins during antenatal life compromises penile growth and inguinoscrotal testicular migration, which are dependent on gonadotropins and androgens. Thus, patients with isolated congenital HH due to deficient hypothalamic GnRH secretion or to pituitary resistance to this neuropeptide, frequently exhibit the cardinal features of antenatal gonadotropin deficiency, namely, cryptorchidism and micropenis. The absence of gonadotropin effects in the pre- and postnatal testes, accounts for the reduction in Sertoli cell mass, small testicular volume, markedly low serum inhibin B levels, and underdeveloped external genitalia frequently exhibited by adults with isolated HH.[236,243] Thus, treatment with GnRH and/or gonadotropins should be primarily addressed to induce virilization and maturation of the seminiferous tubulus and intersticial compartments, which eventually will lead to testicular growth and sperm production.

Considering the evidence on the importance of gonadotropin exposure during the neonatal period for testicular development and maturation, two studies attempted to mimic the neonatal "minipuberty" in patients in whom the diagnosis of HH was established early in infancy[237,239] (Table 2). In both reports, human recombinant gonadotropins were given s.c. during 6 months, time that corresponds to the duration of the minipuberty in males. In both trials, the testicular volume and penile length, as well as serum LH and FSH levels and markers of Sertoli cell stimulation (inhibin B and antimüllerian hormone -AMH-) significantly increased in the treated newborns. Clearly, it remains to be seen if these gonadotropin-treated neonates eventually present a better spermatogenic profile and fertility potential after gonadotropin reexposure in adulthood than those not receiving exogenous gonadotropins early in infancy. If so, then a short, 6-months course of gonadotropin administration might be recommended in newborns with a well established diagnosis of congenital HH in any of its varieties. The development of spermatogenesis in newborns treated with gonadotropins should not be of concern given that during this period the Sertoli cells do not express the androgen receptor, which is necessary for stimulation of spermatogenesis in man.[244] In this vein, it is important to emphasize on the critical role played by surgical correction of cryptorchidism at the appropriate time (usually at 6–12 months of age)[245] in those cases in which gonadotropin therapy was not sufficient to correct this abnormality, which is relatively frequent in congenital HH and that, if untreated, may lead to poor prognosis for fertility in adulthood due to testicular damage.[239,246–250]

Table 2 Studies Using Different Protocols of Exogenous GnRH and/or Gonadotropins as Treatment for HH in Male Individuals.

Study Design	Outcomes/Conclusions	References
Six men with IHH treated with low-dose, pulsatile s.c. GnRH	Long-term episodic GnRH administration can reverse IHH.	[251]
CG administered to 22 men (16 ± 4.2-year-old/ 38.4 ± 30.7 months of treatment) with isolated HH until MTG was achieved	CG, in the absence of exogenous FSH, can complete spermiogenesis in men with partial gonadotropin deficiency. The response to CG in men with isolated HH is predicted by the initial testicular volume.	[247]
Five men with complete isolated HH (30.6 ± 2.2-year-old) treated with pulsatile GnRH in doses of 143–714 ng/kg every 2 h, s.c., while 11 cases (23.1 ± 1.9) received CG (2000 IU) and hMG (75 IU FSH and 75 IU LH) 3 times/week	Pulsatile s.c. GnRH therapy for the first 2 years does not accelerate or enhance TG, hasten the onset of sperm production, or increase sperm output significantly compared to CG/hMG.	[252]
Thirty-eighth men with HH (17 with idiopathic HH, 11 KS, 4 with multiple pituitary hormone deficiencies and 6 with a secondary HH due to surgical removal of a brain tumor) treated with GnRH 2-20 μg i.v. per pulse every 90 min. After GnRH discontinuation, CG treatment was started with 1500–3000 IU i.m. twice weekly	GnRH is a feasible therapy to induce TG as well as spermatogenesis in HH, even in patients in whom this treatment has failed. After GnRH treatment, CG alone can maintain or even improve TD, including spermatogenesis. GnRH treatment may also induce a physiological growth spurt in hypogonadotropic adolescents.	[253]
Sixty men with HH (16 KS, 19 IHH, and 25 with hypopituitarism; 26.3 ± 0.85-year-old). All patients self-administered s.c. injections of FSH (150 IU 3 times/week) and CG (2500 IU 2 times/week) for at least 6 months	s.c. self-administration of highly purified FSH plus CG was well tolerated and effective in stimulating spermatogenesis and steroidogenesis.	[254]
Three prepubertal patients, 2 with idiopathic HH and 1 with acquired HH treated with recFSH s.c. (1.5 IU/kg) 3 times/week for 12 months	In prepubertal boys, recFSH treatment stimulates production of inhibin B, which reflects Sertoli cell function, and induces TG.	[240]

Twenty four patients aged 18–40 years with HH secondary to β-thalassemia major treated with scaled doses of CG alone and thereafter with urinary FSH during 8.6 ± 3.9 years.	Gonadotropin therapy normalized testosterone levels and TV, and induced spermatogenesis. Gonadotropin therapy is an effective replacement therapy for HH due to β-thalassemia major.	[255]
Four men with IHH and 2 KS (30.1 ± 5-year-old) treated with pulsatile GnRH (5–20 μg every 120 min s.c.). Nine cases with IHH and 9 with KS (29.1 ± 7.2-year-old) treated with CG 1000–2500 IU i.m. or s.c. applied twice a week or 75–150 IU hMG applied thrice a week.	GnRH or hCG/hMG are effective therapeutic modalities for patients with IHH or KS.	[256]
Seven prepubertal males with isolated HH (15.4 ± 1.97-year-old) and 7 with PH-associated HH (18.1 ± 3.24-year-old). CG (1000–1500 IU i.m.) and FSH (75-100 IU s.c.) were administered every other day until complete induction of puberty and spermatogenesis were achieved.	Long-term combined CG and FSH therapy is effective in inducing puberty, increasing TV, and stimulating spermatogenesis in adolescent males with isolated HH and PH-associated HH.	[257]
Thirty-seven hypogonadal adolescents were treated with CG/hMG	CG/hMG treatment had a better outcome than testosterone during the induction of puberty, avoiding psychological problems induced by atrophic testes.	[258]
10 men with HH (median age 37 years; 26–48-year-old) requiring induction of spermatogenesis and fertility were treated with CG for 3–6 months, followed by self-administration of recFSH s.c. in combination with CG for 18 months	recFSH is well tolerated and effective in inducing TG, spermatogenesis and fertility in HH men. The efficacy of recFSH seems comparable with urinary FSH at restoring normal fertility in gonadotrophin-deficient men.	[259]
One patient with micropenis treated from 7.9 to 13.7 months of age with recLH and FSH in doses of 20 and 21.3 IU s.c. twice weekly, respectively	Gonadotropin treatment in an infant with HH succeeded in inducing an increase in inhibin B and TG, as well as penile length.	[239]

(Continued)

Table 2 Studies Using Different Protocols of Exogenous GnRH and/or Gonadotropins as Treatment for HH in Male Individuals.—cont'd.

Study Design	Outcomes/Conclusions	References
Twenty-six azoospermatic men (16–48 year-old) with isolated HH pretreated for up to 6 months with CG 2000 IU twice weekly; the dose was titrated after 2 months to maintain serum testosterone within the normal range. Patients then received 18 months of treatment with CG together with recFSH 150 IU 3 times/week, which could be increased to 225 IU after 9 months	recFSH with CG is effective for initiating spermatogenesis in patients with isolated HH, and is well tolerated.	[260]
Pulsatile GnRH therapy during 1–2 years to 76 men with partial or complete IHH. LH, FSH, testosterone, and inhibin B were measured serially and maximum sperm count was recorded.	In complete IHH with absence of puberty at baseline, treatment failed to normalize TV and inhibin B levels. GnRH treatment to the majority of men with IHH and partial or complete puberty normalized LH, FSH, T and inhibin B levels, and also successfully induced spermatogenesis. Favorable predictors of outcome are prior history of sexual maturation, inhibin β levels >60 pg/mL, and absence of cryptorchidism.	[249]
A multicenter, open-label, randomized efficacy and safety study performed with combined CG and recFSH treatment to induce spermatogenesis in HH male patients 30 ± 5.6-year-old	A weekly dose of 450 IU (3 × 150 IU or 2 × 225 IU) recFSH, in addition to CG, was able to induce spermatogenesis in many hypogonadotropic azoospermic men who failed to respond to treatment with CG alone.	[261]
Seventeen adolescents suspected of having HH treated with pulsatile GnRH (20 μg/1.73 m^2 per pulse every 90 min i.v.) for 7 days	Pulsatile GnRH cannot distinguish between a normal DP and a hypothalamic defect in still prepubertal patients. Pulsatile GnRH offers an appropriate way to initiate TG including virilization and fertility in males with HH.	[262]

A 30-year retrospective study involving 36 male patients with primary or secondary HH treated with CG and hMG.	Sperm production in response to long-term treatment with CG/hMG was better in HH patients with initial TV ≥4 mL than in those with smaller (≤4 mL) TVs. Initial TV values in HH subjects are useful to predict the severity and duration of HH as well as the response to gonadotropin therapy.	[248]
Fourteen boys (9.9–17.7-year-old) with prepubertal onset of HH (IHH, $n = 2$; KS, $n = 2$; IPH, $n = 4$; organic PH, $n = 6$) treated with recFSH [3 weekly s.c. doses starting from 1.5 IU/kg (range 180–450 IU/week)] alone (2 months to 2.8 years) prior to induction of puberty with the combination of FSH and CG	recFSH induces prepubertal TG and increases circulating inhibin B levels. Puberty was successfully induced with CG and recFSH following recFSH priming. Inhibin B appears useful for monitoring spermatogenetic activity in boys treated with CG. Despite the extremely small initial TV, 86% primed with recFSH displayed sperm in the ejaculate suggesting beneficial effect of recFSH priming on TF later in life.	[263]
Two neonates (1 with hypopituitarism and 1 with HH) with micropenis and microorchidism treated for 6 months with high doses of recLH and FSH delivered s.c. with an insulin pump	The treatment induced physiological postnatal TG and high-normal activation of Leydig and Sertoli cells.	[237]
A combined analysis of data from four clinical trials conducted in Australia, Europe, Japan, and the United States (100 men, 16–55-year-old with complete idiopathic or acquired HH). Pretreatment with CG for 3–6 months, followed by combination therapy with CG and recFSH (150 IU 3 times weekly) for up to 18 months	recFSH (combined with CG) is effective for the restoration of fertility in the majority of men with HH. Large baseline mean TV, low BMI, and advanced sexual maturity were predictors of good response to therapy. Similar treatment responses were observed across different study populations.	[242]

(Continued)

Table 2 Studies Using Different Protocols of Exogenous GnRH and/or Gonadotropins as Treatment for HH in Male Individuals.—cont'd.

Study Design	Outcomes/Conclusions	References
Nineteen males (14.5–31.0-year-old) with HH, treated with either CG alone (initially at a dose of 500 IU twice weekly, with increments to 1,000 IU at 6 mo and, as puberty progressed, to 1,500 IU 2 times per week) or in combination with recFSH (150–300 IU of 3 times a week, from the fourth month of the higher CG dose), over 6–9 months	Addition of recFSH to CG treatment protocols in adolescent/young adult HH males results in normal TG and may hasten induction of spermatogenesis.	[264]
Retrospective analysis of 10 years hormonal treatment experience of 242 HH patients aged 15 to 52 years and treated with either CG alone, CG plus hMG or testosterone, and followed-up for 6–60 months.	CG or CG/hMG therapy increased TV in 50–75% of patients after 6–18 months of treatment, while <8% did with testosterone. Sperm production was achieved with both CG and CG/hMG treatment only in patients with testicular growth that received gonadotropin therapy. Testosterone treatment did not stimulate testicular growth or spermatogenesis. The study did not allow distinction between primary or secondary HH.	[250]
Eighteen adults (28 ± 2-year-old) with CHH randomly assigned to either 24 months of pulsatile GnRH therapy or 4 months pretreatment with recFSH 75–150 IU s.c. daily (target serum FSH level, 7–9 IU/L), followed by 24 months of pulsatile GnRH	recFSH pretreatment followed by GnRH is successful in inducing TG and fertility in men with CHH with prepubertal testes. recFSH not only appears to maximize the Sertoli cell population but also induces morphologic changes, suggesting broader developmental roles	[265]

Twenty-three men with KS and 12 men with normosmic IHH. Twelve received 8–10 μg of GnRH s.c. every 90 min; 22 received CG i.m.: for the first 3 months 1000 IU injected twice a week and then once every other day for the following 3 months. The dose of CG was increased to 2000 IU after 6 months of treatment and the above cycle was repeated for another 6 months. All patients were treated for 12–14 months and followed-up every 3 months	Boys with HH may be effectively treated with GnRH. The authors suggested that GnRH exhibits higher efficacy than CG for treating adolescent boys with HH	[266]

CG, human chorionic gonadotropin; CHH, Congenital hypogonadotropic hypogonadism; DP, delayed puberty; GnRH, gonadotropin-releasing hormone; hMG, human menopausal gonadotropin; HH, hypogonadotropic hypogonadism; IHH, idiopathic hypogonadotropic hypogonadism; IPH, isolated panhypopituitarism; KS, Kallmann syndrome; MTG, maximal testicular growth; PH, panhypopituitarism; recFSH, recombinant human FSH; TG, testicular growth; TD, testicular development; TF, testicular function; TV, testis volume.

7.2 Gonadotropin Treatment in Adolescents and Adults with HH

Although testosterone administration may induce virilization and masculinization in gonadotropin-deficient adolescent and adult patients with HH, it does not stimulate testicular development.[250,258,267,268] Thus, in order to achieve testicular maturation and stimulate spermatogenesis, it is necessary to consider treatment either with pulsatile GnRH or with exogenous gonadotropins [CG or FSH (or hMG) plus CG] (Table 2). The outcomes of these treatments in terms of testicular maturation and fertility are highly variable and will depend on whether the patient has congenital or acquired HH, with a better prognosis in the latter in terms of fertility potential due to previous exposure to gonadotropins during fetal life, early infancy and/or puberty.[255,257] In those rare cases with documented mutations in *LHB* (see earlier), long-term treatment with CG may apparently be sufficient to promote pubertal development with testicular growth, testosterone production, and in those cases with azoospermia, limited spermatogenesis.[82,86] In the case of *FSHB* mutations, the response to FSH treatment in terms of spermatogenesis is less clear given the absence of data on this issue; nevertheless, administration of FSH to these patients may be, at least theoretically, of benefit for testicular growth and development of spermatogenesis whenever the patient has sufficient production of endogenous testosterone. In fact, there are isolated reports of infertile patients presenting with azoospermia or oligoteratozoospermia and isolated FSH deficiency, but without any evidence of *FSHB* mutations, who were successfully treated with hMG or FSH.[269,270]

Treatment of prepubertal children with congenital HH was initially reported almost two decades ago.[240] Recombinant FSH administration during 1-year stimulated production of inhibin B and stimulated testicular growth, suggesting that FSH is capable of stimulating Sertoli-cell proliferation and elongation of seminiferous tubules, which may potentially increase the sperm-producing capacity later in life.[240] Virilization and testicular maturation can be achieved in adolescents and young adults with partial or complete HH by pulsatile GnRH,[249,251–253,256,262,265,266,271] CG,[247,250,266] or combined CG and FSH (either recombinant or urinary FSH)[242,248,250,252,254,256–261,263,264] administration (Table 2). Although pulsatile GnRH administration, with or without FSH pretreatment[265] seems to be the most physiological treatment and in one recent study exhibited higher efficacy than CG,[266] it is costly and requires long-term use of a portable miniinfusion pump. Barrio and colleagues[257] reported that long-term combined CG and FSH therapy was effective in inducing puberty, increasing testicular volume, and stimulating spermatogenesis in adolescent males with isolated HH and panhypopituitarism-associated HH. Simultaneously, Bouvattier and colleagues[258] reported that treatment with CG and hMG in hypogonadal adolescents had a better outcome than testosterone during the induction of puberty, with less satisfactory results in adolescents with cryptorchidism. One decade later, Raivio and colleagues[263] reported long-term results of treatment with recombinant FSH followed by the combination of FSH and CG in children or adolescents with prepubertal-onset HH. Recombinant FSH stimulated testicular growth and increased serum inhibin B levels appeared useful for monitoring spermatogenic activity. In this study, puberty was successfully induced with CG and FSH following FSH priming.[263] Further, congenital HH was associated with a lower increase in testicular volume and inhibin B levels than postnatally acquired HH, again emphasizing on the important role of the neonatal increase in gonadotropin levels on the testicular response to gonadotropins later life.[263] Once testicular volume, serum testosterone and spermatogenesis has been achieved, the patient can be switched to treatment with either CG alone or testosterone until fertility is desired, when gonadotropin therapy may again be considered if necessary. In fact, it has been shown that switching to CG after induction of spermatogenesis may decrease sperm count but be sufficient to preserve qualitatively normal spermatogenesis.[272] On the other hand, although it has been suggested that the response to subsequent gonadotropin administration is not adversely affected by prior testosterone therapy,[273] a more recent study detected that previous testosterone treatment may be a predictor of slower induction of sperm output.[241] If this is the case, then it might be recommended to induce virilization at puberty with gonadotropins (or GnRH) rather than with testosterone. When treating HH patients, it is important to keep in mind that congenital HH may revert spontaneously in 10–20% of patients[274–276]; thus, a brief withdrawal of treatment with either gonadotropins or testosterone and reassessment of the reproductive axis function is advisable to seek for spontaneous reversion of the hypogonadal state. Since reversal may be temporal in about 5% of these patients,[275,277] long-term monitoring of reproductive function in HH reversal is strongly recommended.

In summary, although virilization of adolescents with HH can be achieved with testosterone administration alone, it is advisable to rather consider gonadotropin therapy to promote testicular growth and maturation, including the development of spermatogenesis, to improve later in life the response to spermatogenesis-inducing gonadotropin therapy when fertility is desired. In adult patients with onset of HH after puberty or in those with isolated HH who desire fertility, gonadotropin therapy is usually effective to promote the reappearance of spermatogenesis and achieve fertility potential. The outcomes of gonadotropin therapy in terms of spermatogenesis and fertility will vary depending on several factors including prior history of undescended testes or spontaneous sexual maturation, testicular size, serum levels of inhibin B, previous treatments received, particular gonadotropin combination employed, and duration of treatment[278] (Table 2).

7.3 FSH Administration in Idiopathic Male Infertility

Idiopathic male infertility (IMI) is a frequent disorder which is usually considered when after a complete diagnostic work up, including clinical, genetic and hormonal testing[279–284] the underlying pathophysiological mechanism remains unknown; these occurs in approximately 30–50% of infertile men.[282,283] In contrast to the well-known effects of gonadotropin therapy in male subjects with HH, the benefit of FSH administration to treat IMI is still uncertain despite the relatively high number of studies reported on this issue.[285–313] Differences among the results of trials involving normogonadotropic men with oligozoospermia, asthenozoospermia, teratozoospermia, or a combination of these semen abnormalities is due to several factors including selection bias, differences in FSH treatment dose and duration,[292,296,305,309] methodological inconsistencies in semen analysis as well as in threshold and normality values, different outcomes considered,[286,290,291,293,295,297,302,303,308,312] variations in inclusion criteria, and absence of adequate controls, to mention the most important.[288,311] A number of positive predictive factors of response to FSH treatment have also been proposed; these include normal serum FSH (<7 IU/L) and inhibin B levels,[304,314–318] absence of maturation abnormalities in germ cells as disclosed by cytological examination of testicular tissue aspirates,[301] relatively high AMH concentrations in seminal plasma,[293] and low sperm DNA fragmentation.[295] Therefore, in well selected cases treatment with FSH may lead to improvement in sperm quality and fertility outcome, particularly when intracytoplasmic-sperm injection is employed.[294]

When considering FSH administration to men with idiopathic infertility associated to abnormal spermiogram, it is important to keep in mind that our knowledge on the exact role of FSH in human spermatogenesis is still incomplete and that it is probably related only to qualitative aspects, more than to full spermatic maturation. In fact, men with loss-of-function mutations in the *FSHR* and mice knockout for either the *Fshr* or *Fshb* and thus, lacking FSH ligand, do not present fertility failure.[19,97,319]

Two recent metaanalysis studies on the effects of FSH treatment in men with idiopathic infertility have been performed.[288,311] The first study, which included only six randomized, controlled studies[288] suggested that in men with idiopathic male factor subfertility a beneficial effect on live birth and pregnancy rate may be achieved with FSH treatment; nevertheless, considering the low number of trials and participants included, the authors acknowledge the limitations of the analysis and concluded that evidence was insufficient to allow final conclusions on this issue.[288] The most recent metaanalysis[311] was more laxed in inclusion criteria than the former and included 15 controlled (randomized or nonrandomized) studies and 1275 infertile men with mild oligozoospermia to severe oligo-asthenoteratozoospermia treated with either recFSH or urinary FSH (614 FSH-treated and 661 placebo-treated or untreated controls). This analysis found that there was a modest, albeit significant improvement in either spontaneous pregnancy rate [OR 4.5 (CI, 2.17–9.33] or pregnancy after assisted reproductive technology (ART) [OR, 1.6 (CI, 1.08–2.37)], independently of the type of FSH preparation employed, cumulative dose of FSH received, or duration of treatment.[311] In the number needed to treat analysis, the study found that 10 men should be treated to achieve one spontaneous pregnancy and 18 men to achieve pregnancy with ART, which actually indicates the relatively low efficacy of the FSH treatment in these cases. It is important to mention that in this metaanalysis, untreated controls exhibited twice higher sperm concentration than the FSH treated men, which may account, at least partially, for the modest results observed. For the secondary outcomes, conservative estimates of overall FSH effects revealed a modest increase in sperm concentration but not in progressive sperm motility or testicular volume; further, there was no correlation between pregnancy rate and any of these secondary outcomes. The finding that FSH dose and treatment duration had no effect on the primary outcome contrasts with at least two placebo-controlled studies[296,309] showing that a high dose of FSH (300 IU every other day) for ≥4–5 months may be more effective than the doses usually employed (75–150 IU, every other day).

In conclusion, we still need to know more precisely the role of FSH and the interplay between this gonadotropin and other testicular paracrine/autocrine factors in male spermatogenesis,[320–322] before assigning a role to exogenous FSH as a therapeutic strategy for IMI. Meanwhile, it is advisable to be more stringent in the criteria for selecting men who should receive FSH treatment for this still poorly defined condition. In this vein, the role of pharmacogenomics in the identification and treatment of men with idiopathic infertility who might respond to exogenous FSH administration is worth mentioning.[80] It has been shown that two SNPs involving the *FSHB* and *FSHR* genes may influence serum concentration of FSH and the response of the FSHR to the gonadotropin stimulus.[80,323–326] As mentioned in Section 2.3, the regulatory SNP within the *FSHB* promoter (–211G/T) has been recognized as a polymorphism that alters the transcriptional activation of *FSHB*[327] leading to significantly reduced serum FSH levels in homozygous male and female T carriers compared with heterozygous and homozygous G carriers.[80,328–332] Whereas in women, the presence of the minor allele does not apparently affect the outcome of ART, in men this particular genetic variant has a major effect since it does not only impact on serum FSH levels but also may associate with lower testicular volume, sperm count, and testosterone levels as well as with higher serum LH concentrations and infertility.[279,328,330–333] In contrast, when the frequent *FSHR* polymorphism 2039A>G [Asn680Ser; in which the Ser680 variant leads to altered FSHR function in vitro][334,335] is considered alone, only homozygous Ser680 women seem to be significantly affected; in fact, this FSHR variant is associated with higher serum FSH levels and lower responses to controlled ovarian hyperstimulation,[80] albeit men carrying this SNP may exhibit lower testicular volumes than carriers of the Asn680 variant.[336] Nevertheless, an interesting interaction between the –211G/T SNP in the *FSHB* and the 2039A>G SNP in the *FSHR* has been described,[279] in which the presence of both the TT variant and the Ser680 variant might represent an unfavorable allele combination in terms of testes volume, serum FSH levels, and fertility potential.[279] In this scenario, oligozoospermic men carrying one or both of these fertility unfavorable variants, may benefit from exogenous FSH treatment, as suggested by at least two studies.[333,337]

8. CONCLUDING REMARKS

In this chapter we attempted to describe how knowledge emerged from basic, pharmacological, and clinical research on gonadotropins has been translated to the clinical arena to treat disease in males with gonadotropin deficiency and other reproductive abnormalities.

A number of highly purified recombinant gonadotropin preparations with high specific activity are currently available to treat reproductive disorders during postnatal life, puberty and adulthood, and new recombinant preparations derived from human cell lines, with carbohydrate structures more similar to those present in the naturally occurring gonadotropins are forthcoming in the very near future.[338] In addition, new strategies to obtain homogenous glycosylation of the gonadotropin subunits are emerging, which could open the possibility to manufacture completely homogeneously glycosylated gonadotropins.[233] Nevertheless, it remains to be established whether these compounds may be superior in improving reproductive outcomes in infertile men than the currently employed preparations. In this vein, the development of synthetic allosteric modulators of the FSHR are of particular interest given their potential to amplify and modulate the FSH stimulus and to provoke preferential effects on specific signaling pathways and/or expression of genes directly involved in germ cell maturation,[339–342] with the advantage of oral administration.

There is little doubt that our knowledge on the physiopathogenesis of IMI is far from complete as it is the role of FSH and autocrine/paracrine factors in regulating sperm maturation and achievement of reproductive competence in man.[8] In this regard, application of "OMIC" techniques may be instrumental to generate new information on these issues and to provide clues for designing new, more effective, and more personalized therapeutic strategies for this frequent disorder.

ACKNOWLEDGMENTS

Research in the senior author's laboratory is supported by grants from the Consejo Nacional de Ciencia y Tecnología (grants 240619 and 252646) and the Universidad Nacional Autónoma de México. Ari Kleinberg, from the Research Support Network, UNAM, kindly performed the graphic design of Fig. 2.

REFERENCES

1. Pierce JG, Parsons TF. Glycoprotein hormones: structure and function. *Annu Rev Biochem.* 1981;50:465–495.
2. Ulloa-Aguirre A, Maldonado A, Damian-Matsumura P, Timossi C. Endocrine regulation of gonadotropin glycosylation. *Arch Med Res.* 2001;32:520–532.
3. Ulloa-Aguirre A, Timossi C, Barrios-de-Tomasi J, Maldonado A, Nayudu P. Impact of carbohydrate heterogeneity in function of follicle-stimulating hormone: studies derived from in vitro and in vivo models. *Biol Reprod.* 2003;69:379–389.
4. Bousfield GR, Dias JA. Synthesis and secretion of gonadotropins including structure-function correlates. *Rev Endocr Metab Disord.* 2011;12:289–302.
5. Ulloa-Aguirre A, Midgley Jr AR, Beitins IZ, Padmanabhan V. Follicle-stimulating isohormones: characterization and physiological relevance. *Endocr Rev.* 1995;16:765–787.

6. Ulloa-Aguirre A, Timossi C. Biochemical and functional aspects of gonadotrophin-releasing hormone and gonadotrophins. *Reprod Biomed Online.* 2000;1:48–62.
7. Bernard DJ, Fortin J, Wang Y, Lamba P. Mechanisms of FSH synthesis: what we know, what we don't, and why you should care. *Fertil Steril.* 2010;93:2465–2485.
8. Huhtaniemi I. A short evolutionary history of FSH-stimulated spermatogenesis. *Hormones.* 2015;14:468–478.
9. Richards JS, Pangas SA. The ovary: basic biology and clinical implications. *J Clin Invest.* 2010;120:963–972.
10. Bukovsky A, Indrapichate K, Fujiwara H, et al. Multiple luteinizing hormone receptor (LHR) protein variants, interspecies reactivity of anti-LHR mAb clone 3B5, subcellular localization of LHR in human placenta, pelvic floor and brain, and possible role for LHR in the development of abnormal pregnancy, pelvic floor disorders and Alzheimer's disease. *Reprod Biol Endocrinol.* 2003;1:46.
11. Kumar TR. Extragonadal FSH receptor: is it real? *Biol Reprod.* 2014;91:99.
12. Banerjee P, Fazleabas AT. Extragonadal actions of chorionic gonadotropin. *Rev Endocr Metab Disord.* 2011;12:323–332.
13. Cole LA. Biological functions of hCG and hCG-related molecules. *Reprod Biol Endocrinol.* 2010;8:102.
14. Cole LA. Hyperglycosylated hCG, a review. *Placenta.* 2010;31:653–664.
15. Fournier T, Guibourdenche J, Evain-Brion D. Review: hCGs: different sources of production, different glycoforms and functions. *Placenta.* 2015;36(Suppl 1):S60–S65.
16. Tarulli GA, Stanton PG, Meachem SJ. Is the adult Sertoli cell terminally differentiated? *Biol Reprod.* 2012;87:13.
17. Walker WH, Cheng J. FSH and testosterone signaling in Sertoli cells. *Reproduction.* 2005;130:15–28.
18. Sharpe RM, McKinnell C, Kivlin C, Fisher JS. Proliferation and functional maturation of Sertoli cells, and their relevance to disorders of testis function in adulthood. *Reproduction.* 2003;125:769–784.
19. Tapanainen JS, Aittomaki K, Min J, Vaskivuo T, Huhtaniemi IT. Men homozygous for an inactivating mutation of the follicle-stimulating hormone (FSH) receptor gene present variable suppression of spermatogenesis and fertility. *Nat Genet.* 1997;15: 205–206.
20. Matsumoto AM, Karpas AE, Paulsen CA, Bremner WJ. Reinitiation of sperm production in gonadotropin-suppressed normal men by administration of follicle-stimulating hormone. *J Clin Invest.* 1983;72:1005–1015.
21. Matsumoto AM, Paulsen CA, Bremner WJ. Stimulation of sperm production by human luteinizing hormone in gonadotropin-suppressed normal men. *J Clin Endocrinol Metab.* 1984;59:882–887.
22. Shiraishi K, Ascoli M. Lutropin/choriogonadotropin stimulate the proliferation of primary cultures of rat Leydig cells through a pathway that involves activation of the extracellularly regulated kinase 1/2 cascade. *Endocrinology.* 2007;148: 3214–3225.
23. Tai P, Shiraishi K, Ascoli M. Activation of the lutropin/choriogonadotropin receptor inhibits apoptosis of immature Leydig cells in primary culture. *Endocrinology.* 2009;150:3766–3773.
24. Miller WL, Auchus RJ. The molecular biology, biochemistry, and physiology of human steroidogenesis and its disorders. *Endocr Rev.* 2011;32:81–151.
25. Gregory SJ, Kaiser UB. Regulation of gonadotropins by inhibin and activin. *Semin Reprod Med.* 2004;22:253–267.
26. Fiddes JC, Talmadge K. Structure, expression, and evolution of the genes for the human glycoprotein hormones. *Recent Prog Horm Res.* 1984;40:43–78.

27. Naylor SL, Chin WW, Goodman HM, Lalley PA, Grzeschik KH, Sakaguchi AY. Chromosome assignment of genes encoding the alpha and beta subunits of glycoprotein hormones in man and mouse. *Somatic Cell Genet.* 1983;9:757–770.
28. Gharib SD, Wierman ME, Shupnik MA, Chin WW. Molecular biology of the pituitary gonadotropins. *Endocr Rev.* 1990;11:177–199.
29. Jameson JL, Becker CB, Lindell CM, Habener JF. Human follicle-stimulating hormone beta-subunit gene encodes multiple messenger ribonucleic acids. *Mol Endocrinol.* 1988;2:806–815.
30. Nagirnaja L, Rull K, Uuskula L, Hallast P, Grigorova M, Laan M. Genomics and genetics of gonadotropin beta-subunit genes: Unique FSHB and duplicated LHB/CGB loci. *Mol Cell Endocrinol.* 2010;329:4–16.
31. Rull K, Laan M. Expression of beta-subunit of HCG genes during normal and failed pregnancy. *Hum Reprod.* 2005;20:3360–3368.
32. Cocquebert M, Berndt S, Segond N, et al. Comparative expression of hCG beta-genes in human trophoblast from early and late first-trimester placentas. *Am J Physiol Endocrinol Metab.* 2012;303:E950–E958.
33. Jiang X, Dias JA, He X. Structural biology of glycoprotein hormones and their receptors: insights to signaling. *Mol Cell Endocrinol.* 2014;382:424–451.
34. Fox KM, Dias JA, Van Roey P. Three-dimensional structure of human follicle-stimulating hormone. *Mol Endocrinol.* 2001;15:378–389.
35. Lapthorn AJ, Harris DC, Littlejohn A, et al. Crystal structure of human chorionic gonadotropin. *Nature.* 1994;369:455–461.
36. Fan QR, Hendrickson WA. Structure of human follicle-stimulating hormone in complex with its receptor. *Nature.* 2005;433:269–277.
37. Bousfield GR, Baker VL, Gotschall RR, Butnev VY. Carbohydrate analysis of glycoprotein hormones. *Methods.* 2000;21:15–39.
38. Matzuk MM, Boime I. Site-specific mutagenesis defines the intracellular role of the asparagine-linked oligosaccharides of chorionic gonadotropin beta subunit. *J Biol Chem.* 1988;263:17106–17111.
39. Matzuk MM, Boime I. The role of the asparagine-linked oligosaccharides of the alpha subunit in the secretion and assembly of human chorionic gonadotrophin. *J Cell Biol.* 1988;106:1049–1059.
40. Bielinska M, Matzuk MM, Boime I. Site-specific processing of the N-linked oligosaccharides of the human chorionic gonadotropin alpha subunit. *J Biol Chem.* 1989;264:17113–17118.
41. Matzuk MM, Hsueh AJ, Lapolt P, Tsafriri A, Keene JL, Boime I. The biological role of the carboxyl-terminal extension of human chorionic gonadotropin [corrected] beta-subunit. *Endocrinology.* 1990;126:376–383.
42. Muyan M, Ruddon RW, Norton SE, Boime I, Bedows E. Dissociation of early folding events from assembly of the human lutropin beta-subunit. *Mol Endocrinol.* 1998;12:1640–1649.
43. Sairam MR. Role of carbohydrates in glycoprotein hormone signal transduction. *Faseb J.* 1989;3:1915–1926.
44. Sairam MR, Bhargavi GN. A role for glycosylation of the alpha subunit in transduction of biological signal in glycoprotein hormones. *Science.* 1985;229:65–67.
45. Smith PL, Kaetzel D, Nilson J, Baenziger JU. The sialylated oligosaccharides of recombinant bovine lutropin modulate hormone bioactivity. *J Biol Chem.* 1990;265:874–881.
46. Wehbi V, Tranchant T, Durand G, et al. Partially deglycosylated equine LH preferentially activates beta-arrestin-dependent signaling at the follicle-stimulating hormone receptor. *Mol Endocrinol.* 2010;24:561–573.

47. Jiang C, Hou X, Wang C, et al. Hypoglycosylated hFSH has Greater Bioactivity than fully glycosylated recombinant hFSH in human granulosa cells. *J Clin Endocrinol Metab.* 2015;100:E852–E860.
48. Meher BR, Dixit A, Bousfield GR, Lushington GH. Glycosylation effects on FSH-FSHR interaction dynamics: a case study of different FSH glycoforms by molecular dynamics simulations. *PLoS One.* 2015;10:e0137897.
49. Bishop LA, Nguyen TV, Schofield PR. Both of the beta-subunit carbohydrate residues of follicle-stimulating hormone determine the metabolic clearance rate and in vivo potency. *Endocrinology.* 1995;136:2635–2640.
50. Smith PL, Baenziger JU. A pituitary N-acetylgalactosamine transferase that specifically recognizes glycoprotein hormones. *Science.* 1988;242:930–933.
51. Smith PL, Baenziger JU. Molecular basis of recognition by the glycoprotein hormone-specific N-acetylgalactosamine-transferase. *Proc Natl Acad Sci USA.* 1992;89: 329–333.
52. Weisshaar G, Hiyama J, Renwick AG. Site-specific N-glycosylation of human chorionic gonadotrophin–structural analysis of glycopeptides by one- and two-dimensional 1H NMR spectroscopy. *Glycobiology.* 1991;1:393–404.
53. Dalpathado DS, Irungu J, Go EP, et al. Comparative glycomics of the glycoprotein follicle stimulating hormone: glycopeptide analysis of isolates from two mammalian species. *Biochemistry.* 2006;45:8665–8673.
54. Dharmesh SM, Baenziger JU. Estrogen modulates expression of the glycosyltransferases that synthesize sulfated oligosaccharides on lutropin. *Proc Natl Acad Sci USA.* 1993;90:11127–11131.
55. Olivares A, Cardenas M, Timossi C, Zarinan T, Diaz-Sanchez V, Ulloa-Aguirre A. Reactivity of different LH and FSH standards and preparations in the world health organization matched reagents for enzyme-linked immunoassays of gonadotrophins. *Hum Reprod.* 2000;15:2285–2291.
56. Wide L, Eriksson K, Sluss PM, Hall JE. Serum half-life of pituitary gonadotropins is decreased by sulfonation and increased by sialylation in women. *J Clin Endocrinol Metab.* 2009;94:958–964.
57. Jiang X, Fischer D, Chen X, et al. Evidence for follicle-stimulating hormone receptor as a functional trimer. *J Biol Chem.* 2014;289:14273–14282.
58. Bousfield GR, Butnev VY, Butnev VY, et al. Differential effects of alpha subunit Asparagine56 oligosaccharide structure on equine lutropin and follitropin hybrid conformation and receptor-binding activity. *Biochemistry.* 2004;43:10817–10833.
59. Matzuk MM, Keene JL, Boime I. Site specificity of the chorionic gonadotropin N-linked oligosaccharides in signal transduction. *J Biol Chem.* 1989;264:2409–2414.
60. Valove FM, Finch C, Anasti JN, Froehlich J, Flack MR. Receptor binding and signal transduction are dissociable functions requiring different sites on follicle-stimulating hormone. *Endocrinology.* 1994;135:2657–2661.
61. Bishop LA, Robertson DM, Cahir N, Schofield PR. Specific roles for the asparagine-linked carbohydrate residues of recombinant human follicle stimulating hormone in receptor binding and signal transduction. *Mol Endocrinol.* 1994;8:722–731.
62. Flack MR, Froehlich J, Bennet AP, Anasti J, Nisula BC. Site-directed mutagenesis defines the individual roles of the glycosylation sites on follicle-stimulating hormone. *J Biol Chem.* 1994;269:14015–14020.
63. Bousfield GR, Butnev VY, Rueda-Santos MA, Brown A, Hall AS, Harvey DJ. Macro- and micro-heterogeneity in pituitary and urinary follicle-stimulating hormone glycosylation. *J Glycomics Lipidomics.* 2014;4:125.
64. Ulloa-Aguirre A, Espinoza R, Damian-Matsumura P, et al. Studies on the microheterogeneity of anterior pituitary follicle-stimulating hormone in the female rat. Isoelectric focusing pattern throughout the estrous cycle. *Biol Reprod.* 1988;38:70–78.

65. Walton WJ, Nguyen VT, Butnev VY, Singh V, Moore WT, Bousfield GR. Characterization of human FSH isoforms reveals a nonglycosylated beta-subunit in addition to the conventional glycosylated beta-subunit. *J Clin Endocrinol Metab.* 2001;86:3675–3685.
66. Bousfield GR, Butnev VY, Walton WJ, et al. All-or-none N-glycosylation in primate follicle-stimulating hormone beta-subunits. *Mol Cell Endocrinol.* 2007;260–262:40–48.
67. Bousfield GR, Butnev VY, Butnev VY, Hiromasa Y, Harvey DJ, May JV. Hypo-glycosylated human follicle-stimulating hormone (hFSH(21/18)) is much more active in vitro than fully-glycosylated hFSH (hFSH(24)). *Mol Cell Endocrinol.* 2014;382: 989–997.
68. Wide L, Eriksson K. Dynamic changes in glycosylation and glycan composition of serum FSH and LH during natural ovarian stimulation. *Ups J Med Sci.* 2013;118: 153–164.
69. Wide L, Wide M. Higher plasma disappearance rate in the mouse for pituitary follicle-stimulating hormone of young women compared to that of men and elderly women. *J Clin Endocrinol Metab.* 1984;58:426–429.
70. Furui K, Suganuma N, Tsukahara S, et al. Identification of two point mutations in the gene coding luteinizing hormone (LH) beta-subunit, associated with immunologically anomalous LH variants. *J Clin Endocrinol Metab.* 1994;78:107–113.
71. Okuda K, Takamatsu J, Okazaki T, Yamada T, Saeki M, Sugimoto O. Hereditary abnormality of luteinizing hormone resulting in discrepant serum concentrations determined by different assays. *Endocr J.* 1994;41:639–644.
72. Takahashi K, Kurioka H, Ozaki T, et al. Increased prevalence of luteinizing hormone beta-subunit variant in Japanese infertility patients. *Hum Reprod.* 1998;13:3338–3344.
73. Takahashi K, Ozaki T, Okada M, Kurioka H, Kanasaki H, Miyazaki K. Increased prevalence of luteinizing hormone beta-subunit variant in patients with premature ovarian failure. *Fertil Steril.* 1999;71:96–101.
74. Raivio T, Huhtaniemi I, Anttila R, et al. The role of luteinizing hormone-beta gene polymorphism in the onset and progression of puberty in healthy boys. *J Clin Endocrinol Metab.* 1996;81:3278–3282.
75. Lamminen T, Huhtaniemi I. A common genetic variant of luteinizing hormone; relation to normal and aberrant pituitary-gonadal function. *Eur J Pharmacol.* 2001;414:1–7.
76. Lamminen T, Jiang M, Manna PR, et al. Functional study of a recombinant form of human LHbeta-subunit variant carrying the Gly(102)Ser mutation found in Asian populations. *Mol Hum Reprod.* 2002;8:887–892.
77. Ramanujam LN, Liao WX, Roy AC, Ng SC. Association of molecular variants of luteinizing hormone with male infertility. *Hum Reprod.* 2000;15:925–928.
78. Ramanujam LN, Liao WX, Roy AC, Loganath A, Goh HH, Ng SC. Association of molecular variants of luteinizing hormone with menstrual disorders. *Clin Endocrinol.* 1999;51:243–246.
79. Jiang M, Lamminen T, Pakarinen P, et al. A novel Ala(-3)Thr mutation in the signal peptide of human luteinizing hormone beta-subunit: potentiation of the inositol phosphate signalling pathway and attenuation of the adenylate cyclase pathway by recombinant variant hormone. *Mol Hum Reprod.* 2002;8:201–212.
80. Simoni M, Casarini L. Mechanisms in endocrinology: genetics of FSH action: a 2014-and-beyond view. *Eur J Endocrinol.* 2014;170:R91–R107.
81. Achard C, Courtillot C, Lahuna O, et al. Normal spermatogenesis in a man with mutant luteinizing hormone. *N Engl J Med.* 2009;361:1856–1863.
82. Basciani S, Watanabe M, Mariani S, et al. Hypogonadism in a patient with two novel mutations of the luteinizing hormone beta-subunit gene expressed in a compound heterozygous form. *J Clin Endocrinol Metab.* 2012;97:3031–3038.

83. Lofrano-Porto A, Barra GB, Giacomini LA, et al. Luteinizing hormone beta mutation and hypogonadism in men and women. *N Engl J Med.* 2007;357:897–904.
84. Valdes-Socin H, Salvi R, Daly AF, et al. Hypogonadism in a patient with a mutation in the luteinizing hormone beta-subunit gene. *N Engl J Med.* 2004;351:2619–2625.
85. Valdes-Socin H, Salvi R, Thiry A, et al. Testicular effects of isolated luteinizing hormone deficiency and reversal by long-term human chorionic gonadotropin treatment. *J Clin Endocrinol Metab.* 2009;94:3–4.
86. Weiss J, Axelrod L, Whitcomb RW, Harris PE, Crowley WF, Jameson JL. Hypogonadism caused by a single amino acid substitution in the beta subunit of luteinizing hormone. *N Engl J Med.* 1992;326:179–183.
87. Arnhold IJ, Lofrano-Porto A, Latronico AC. Inactivating mutations of luteinizing hormone beta-subunit or luteinizing hormone receptor cause oligo-amenorrhea and infertility in women. *Horm Res.* 2009;71:75–82.
88. Layman LC, Porto AL, Xie J, et al. FSH beta gene mutations in a female with partial breast development and a male sibling with normal puberty and azoospermia. *J Clin Endocrinol Metab.* 2002;87:3702–3707.
89. Lindstedt G, Nystrom E, Matthews C, Ernest I, Janson PO, Chatterjee K. Follitropin (FSH) deficiency in an infertile male due to FSHbeta gene mutation. A syndrome of normal puberty and virilization but underdeveloped testicles with azoospermia, low FSH but high lutropin and normal serum testosterone concentrations. *Clin Chem Lab Med.* 1998;36:663–665.
90. Phillip M, Arbelle JE, Segev Y, Parvari R. Male hypogonadism due to a mutation in the gene for the beta-subunit of follicle-stimulating hormone. *N Engl J Med.* 1998;338: 1729–1732.
91. Berger K, Souza H, Brito VN, d'Alva CB, Mendonca BB, Latronico AC. Clinical and hormonal features of selective follicle-stimulating hormone (FSH) deficiency due to FSH beta-subunit gene mutations in both sexes. *Fertil Steril.* 2005;83:466–470.
92. Kottler ML, Chou YY, Chabre O, et al. A new FSHbeta mutation in a 29-year-old woman with primary amenorrhea and isolated FSH deficiency: functional characterization and ovarian response to human recombinant FSH. *Eur J Endocrinol.* 2010;162:633–641.
93. Layman LC, Lee EJ, Peak DB, et al. Delayed puberty and hypogonadism caused by mutations in the follicle-stimulating hormone beta-subunit gene. *N Engl J Med.* 1997;337:607–611.
94. Matthews C, Chatterjee VK. Isolated deficiency of follicle-stimulating hormone re-revisited. *N Engl J Med.* 1997;337:642.
95. Matthews CH, Borgato S, Beck-Peccoz P, et al. Primary amenorrhoea and infertility due to a mutation in the beta-subunit of follicle-stimulating hormone. *Nat Genet.* 1993;5:83–86.
96. Kumar TR, Palapattu G, Wang P, et al. Transgenic models to study gonadotropin function: the role of follicle-stimulating hormone in gonadal growth and tumorigenesis. *Mol Endocrinol.* 1999;13:851–865.
97. Kumar TR, Wang Y, Lu N, Matzuk MM. Follicle stimulating hormone is required for ovarian follicle maturation but not male fertility. *Nat Genet.* 1997;15:201–204.
98. Layman LC, Cohen DP, Jin M, et al. Mutations in gonadotropin-releasing hormone receptor gene cause hypogonadotropic hypogonadism. *Nat Genet.* 1998;18: 14–15.
99. Simsek E, Montenegro LR, Binay C, Demiral M, Acikalin MF, Latronico AC. Clinical and hormonal features of a male adolescent with congenital isolated follicle-stimulating hormone deficiency. *Horm Res Paediatr.* 2016;85:207–212.
100. Knobil E. The neuroendocrine control of the menstrual cycle. *Recent Prog Horm Res.* 1980;36:53–88.

101. Wildt L, Hausler A, Marshall G, et al. Frequency and amplitude of gonadotropin-releasing hormone stimulation and gonadotropin secretion in the rhesus monkey. *Endocrinology.* 1981;109:376–385.
102. Marshall JC, Dalkin AC, Haisenleder DJ, Paul SJ, Ortolano GA, Kelch RP. Gonadotropin-releasing hormone pulses: regulators of gonadotropin synthesis and ovulatory cycles. *Recent Prog Horm Res.* 1991;47:155–187.
103. Skorupskaite K, George JT, Anderson RA. The kisspeptin-GnRH pathway in human reproductive health and disease. *Hum Reprod Update.* 2014;20:485–500.
104. Ciccone NA, Kaiser UB. The biology of gonadotroph regulation. *Curr Opin Endocrinol Diabetes Obes.* 2009;16:321–327.
105. Grosse R, Schmid A, Schoneberg T, et al. Gonadotropin-releasing hormone receptor initiates multiple signaling pathways by exclusively coupling to G(q/11) proteins. *J Biol Chem.* 2000;275:9193–9200.
106. Stanislaus D, Ponder S, Ji TH, Conn PM. Gonadotropin-releasing hormone receptor couples to multiple G proteins in rat gonadotrophs and in GGH3 cells: evidence from palmitoylation and overexpression of G proteins. *Biol Reprod.* 1998;59:579–586.
107. Stanislaus D, Janovick JA, Brothers S, Conn PM. Regulation of G(q/11)alpha by the gonadotropin-releasing hormone receptor. *Mol Endocrinol.* 1997;11:738–746.
108. Sviridonov L, Dobkin-Bekman M, Shterntal B, et al. Differential signaling of the GnRH receptor in pituitary gonadotrope cell lines and prostate cancer cell lines. *Mol Cell Endocrinol.* 2013;369:107–118.
109. Tsutsumi R, Mistry D, Webster NJ. Signaling responses to pulsatile gonadotropin-releasing hormone in LbetaT2 gonadotrope cells. *J Biol Chem.* 2010;285: 20262–20272.
110. Liu F, Usui I, Evans LG, et al. Involvement of both G(q/11) and G(s) proteins in gonadotropin-releasing hormone receptor-mediated signaling in L beta T2 cells. *J Biol Chem.* 2002;277:32099–32108.
111. Han XB, Conn PM. The role of protein kinases A and C pathways in the regulation of mitogen-activated protein kinase activation in response to gonadotropin-releasing hormone receptor activation. *Endocrinology.* 1999;140:2241–2251.
112. Kraus S, Naor Z, Seger R. Intracellular signaling pathways mediated by the gonadotropin-releasing hormone (GnRH) receptor. *Arch Med Res.* 2001;32:499–509.
113. Naor Z, Huhtaniemi I. Interactions of the GnRH receptor with heterotrimeric G proteins. *Front Neuroendocrinol.* 2013;34:88–94.
114. Bar-Lev TH, Harris D, Tomic M, et al. Role of PI4K and PI3K-AKT in ERK1/2 activation by GnRH in the pituitary gonadotropes. *Mol Cell Endocrinol.* 2015;415:12–23.
115. Dobkin-Bekman M, Naidich M, Pawson AJ, Millar RP, Seger R, Naor Z. Activation of mitogen-activated protein kinase (MAPK) by GnRH is cell-context dependent. *Mol Cell Endocrinol.* 2006;252:184–190.
116. Harris D, Bonfil D, Chuderland D, Kraus S, Seger R, Naor Z. Activation of MAPK cascades by GnRH: ERK and Jun N-terminal kinase are involved in basal and GnRH-stimulated activity of the glycoprotein hormone LHbeta-subunit promoter. *Endocrinology.* 2002;143:1018–1025.
117. Ferris HA, Shupnik MA. Mechanisms for pulsatile regulation of the gonadotropin subunit genes by GNRH1. *Biol Reprod.* 2006;74:993–998.
118. Shupnik MA. Effects of gonadotropin-releasing hormone on rat gonadotropin gene transcription in vitro: requirement for pulsatile administration for luteinizing hormone-beta gene stimulation. *Mol Endocrinol.* 1990;4:1444–1450.
119. Kanasaki H, Bedecarrats GY, Kam KY, Xu S, Kaiser UB. Gonadotropin-releasing hormone pulse frequency-dependent activation of extracellular signal-regulated kinase pathways in perifused LbetaT2 cells. *Endocrinology.* 2005;146:5503–5513.

120. Naor Z. Signaling by G-protein-coupled receptor (GPCR): studies on the GnRH receptor. *Front Neuroendocrinol.* 2009;30:10–29.
121. Zhang T, Roberson MS. Role of MAP kinase phosphatases in GnRH-dependent activation of MAP kinases. *J Mol Endocrinol.* 2006;36:41–50.
122. Purwana IN, Kanasaki H, Mijiddorj T, Oride A, Miyazaki K. Induction of dual-specificity phosphatase 1 (DUSP1) by pulsatile gonadotropin-releasing hormone stimulation: role for gonadotropin subunit expression in mouse pituitary LbetaT2 cells. *Biol Reprod.* 2011;84:996–1004.
123. Tsaneva-Atanasova K, Caunt CJ, Armstrong SP, Perrett RM, McArdle CA. Decoding neurohormone pulse frequency by convergent signalling modules. *Biochem Soc Trans.* 2012;40:273–278.
124. Armstrong SP, Caunt CJ, Fowkes RC, Tsaneva-Atanasova K, McArdle CA. Pulsatile and sustained gonadotropin-releasing hormone (GnRH) receptor signaling: does the ERK signaling pathway decode GnRH pulse frequency? *J Biol Chem.* 2010;285: 24360–24371.
125. Armstrong SP, Caunt CJ, Fowkes RC, Tsaneva-Atanasova K, McArdle CA. Pulsatile and sustained gonadotropin-releasing hormone (GnRH) receptor signaling: does the Ca2+/NFAT signaling pathway decode GnRH pulse frequency? *J Biol Chem.* 2009;284:35746–35757.
126. Armstrong SP, Caunt CJ, McArdle CA. Gonadotropin-releasing hormone and protein kinase C signaling to ERK: spatiotemporal regulation of ERK by docking domains and dual-specificity phosphatases. *Mol Endocrinol.* 2009;23:510–519.
127. Salisbury TB, Binder AK, Nilson JH. Welcoming beta-catenin to the gonadotropin-releasing hormone transcriptional network in gonadotropes. *Mol Endocrinol.* 2008;22: 1295–1303.
128. Haisenleder DJ, Dalkin AC, Ortolano GA, Marshall JC, Shupnik MA. A pulsatile gonadotropin-releasing hormone stimulus is required to increase transcription of the gonadotropin subunit genes: evidence for differential regulation of transcription by pulse frequency in vivo. *Endocrinology.* 1991;128:509–517.
129. Shupnik MA. Gonadotropin gene modulation by steroids and gonadotropin-releasing hormone. *Biol Reprod.* 1996;54:279–286.
130. Heckert LL, Wilson EM, Nilson JH. Transcriptional repression of the alpha-subunit gene by androgen receptor occurs independently of DNA binding but requires the DNA-binding and ligand-binding domains of the receptor. *Mol Endocrinol.* 1997;11:1497–1506.
131. Jameson JL, Jaffe RC, Gleason SL, Habener JF. Transcriptional regulation of chorionic gonadotropin alpha- and beta-subunit gene expression by 8-bromo-adenosine 3',5'-monophosphate. *Endocrinology.* 1986;119:2560–2567.
132. Quirk CC, Lozada KL, Keri RA, Nilson JH. A single Pitx1 binding site is essential for activity of the LHbeta promoter in transgenic mice. *Mol Endocrinol.* 2001;15: 734–746.
133. Jorgensen JS, Quirk CC, Nilson JH. Multiple and overlapping combinatorial codes orchestrate hormonal responsiveness and dictate cell-specific expression of the genes encoding luteinizing hormone. *Endocr Rev.* 2004;25:521–542.
134. Barnhart KM, Mellon PL. The orphan nuclear receptor, steroidogenic factor-1, regulates the glycoprotein hormone alpha-subunit gene in pituitary gonadotropes. *Mol Endocrinol.* 1994;8:878–885.
135. Horn F, Windle JJ, Barnhart KM, Mellon PL. Tissue-specific gene expression in the pituitary: the glycoprotein hormone alpha-subunit gene is regulated by a gonadotrope-specific protein. *Mol Cell Biol.* 1992;12:2143–2153.
136. Ingraham HA, Lala DS, Ikeda Y, et al. The nuclear receptor steroidogenic factor 1 acts at multiple levels of the reproductive axis. *Genes Dev.* 1994;8:2302–2312.

137. Heckert LL, Schultz K, Nilson JH. Different composite regulatory elements direct expression of the human alpha subunit gene to pituitary and placenta. *J Biol Chem.* 1995;270:26497–26504.
138. Lawson MA, Tsutsumi R, Zhang H, et al. Pulse sensitivity of the luteinizing hormone beta promoter is determined by a negative feedback loop Involving early growth response-1 and Ngfi-A binding protein 1 and 2. *Mol Endocrinol.* 2007;21: 1175–1191.
139. Mouillet JF, Sonnenberg-Hirche C, Yan X, Sadovsky Y. p300 regulates the synergy of steroidogenic factor-1 and early growth response-1 in activating luteinizing hormone-beta subunit gene. *J Biol Chem.* 2004;279:7832–7839.
140. Mistry DS, Tsutsumi R, Fernandez M, et al. Gonadotropin-releasing hormone pulse sensitivity of follicle-stimulating hormone-beta gene is mediated by differential expression of positive regulatory activator protein 1 factors and corepressors SKIL and TGIF1. *Mol Endocrinol.* 2011;25:1387–1403.
141. Lim S, Luo M, Koh M, et al. Distinct mechanisms involving diverse histone deacetylases repress expression of the two gonadotropin beta-subunit genes in immature gonadotropes, and their actions are overcome by gonadotropin-releasing hormone. *Mol Cell Biol.* 2007;27:4105–4120.
142. Wurmbach E, Yuen T, Ebersole BJ, Sealfon SC. Gonadotropin-releasing hormone receptor-coupled gene network organization. *J Biol Chem.* 2001;276:47195–47201.
143. Shupnik MA, Fallest PC. Pulsatile GnRH regulation of gonadotropin subunit gene transcription. *Neurosci Biobehav Rev.* 1994;18:597–599.
144. Garcia-Galiano D, Pinilla L, Tena-Sempere M. Sex steroids and the control of the Kiss1 system: developmental roles and major regulatory actions. *J Neuroendocrinol.* 2012;24: 22–33.
145. Shupnik MA, Rosenzweig BA. Identification of an estrogen-responsive element in the rat LH beta gene. DNA-estrogen receptor interactions and functional analysis. *J Biol Chem.* 1991;266:17084–17091.
146. Luo M, Koh M, Feng J, Wu Q, Melamed P. Cross talk in hormonally regulated gene transcription through induction of estrogen receptor ubiquitylation. *Mol Cell Biol.* 2005;25:7386–7398.
147. Kowase T, Walsh HE, Darling DS, Shupnik MA. Estrogen enhances gonadotropin-releasing hormone-stimulated transcription of the luteinizing hormone subunit promoters via altered expression of stimulatory and suppressive transcription factors. *Endocrinology.* 2007;148:6083–6091.
148. Thackray VG, Hunnicutt JL, Memon AK, Ghochani Y, Mellon PL. Progesterone Inhibits basal and gonadotropin-releasing hormone induction of luteinizing hormone beta-subunit gene expression. *Endocrinology.* 2009;150:2395–2403.
149. Thackray VG, Mellon PL, Coss D. Hormones in synergy: regulation of the pituitary gonadotropin genes. *Mol Cell Endocrinol.* 2010;314:192–203.
150. Thackray VG, McGillivray SM, Mellon PL. Androgens, progestins, and glucocorticoids induce follicle-stimulating hormone beta-subunit gene expression at the level of the gonadotrope. *Mol Endocrinol.* 2006;20:2062–2079.
151. Spady TJ, Shayya R, Thackray VG, Ehrensberger L, Bailey JS, Mellon PL. Androgen regulates follicle-stimulating hormone beta gene expression in an activin-dependent manner in immortalized gonadotropes. *Mol Endocrinol.* 2004;18:925–940.
152. Issad T, Strobel A, Camoin L, Ozata M, Strosberg AD. Leptin and puberty in humans: hypothesis of the critical adipose mass revisited. *Diabetes Metab.* 1998;24:376–378.
153. Strobel A, Issad T, Camoin L, Ozata M, Strosberg AD. A leptin missense mutation associated with hypogonadism and morbid obesity. *Nat Genet.* 1998;18:213–215.
154. Clement K, Vaisse C, Lahlou N, et al. A mutation in the human leptin receptor gene causes obesity and pituitary dysfunction. *Nature.* 1998;392:398–401.

155. Wu Y, Luo H, Liu J, Kang D, McNeilly AS, Cui S. LIM homeodomain transcription factor Isl-1 enhances follicle stimulating hormone-beta and luteinizing hormone-beta gene expression and mediates the activation of leptin on gonadotropin synthesis. *Endocrinology.* 2010;151:4787–4800.
156. Nicol L, Faure MO, McNeilly JR, Fontaine J, Taragnat C, McNeilly AS. Bone morphogenetic protein-4 interacts with activin and GnRH to modulate gonadotrophin secretion in LbetaT2 gonadotrophs. *J Endocrinol.* 2008;196:497–507.
157. Takeda M, Otsuka F, Takahashi H, et al. Interaction between gonadotropin-releasing hormone and bone morphogenetic protein-6 and -7 signaling in LbetaT2 gonadotrope cells. *Mol Cell Endocrinol.* 2012;348:147–154.
158. Zawel L, Dai JL, Buckhaults P, et al. Human Smad3 and Smad4 are sequence-specific transcription activators. *Mol Cell.* 1998;1:611–617.
159. Lamba P, Fortin J, Tran S, Wang Y, Bernard DJ. A novel role for the forkhead transcription factor FOXL2 in activin A-regulated follicle-stimulating hormone beta subunit transcription. *Mol Endocrinol.* 2009;23:1001–1013.
160. Bernard DJ, Tran S. Mechanisms of activin-stimulated FSH synthesis: the story of a pig and a FOX. *Biol Reprod.* 2013;88:78.
161. Tran S, Zhou X, Lafleur C, et al. Impaired fertility and FSH synthesis in gonadotrope-specific Foxl2 knockout mice. *Mol Endocrinol.* 2013;27:407–421.
162. Blount AL, Schmidt K, Justice NJ, Vale WW, Fischer WH, Bilezikjian LM. FoxL2 and Smad3 coordinately regulate follistatin gene transcription. *J Biol Chem.* 2009;284: 7631–7645.
163. Choi SG, Wang Q, Jia J, Pincas H, Turgeon JL, Sealfon SC. Growth differentiation factor 9 (GDF9) forms an incoherent feed-forward loop modulating follicle-stimulating hormone beta-subunit (FSHbeta) gene expression. *J Biol Chem.* 2014;289: 16164–16175.
164. Pincas H, Choi SG, Wang Q, Jia J, Turgeon JL, Sealfon SC. Outside the box signaling: secreted factors modulate GnRH receptor-mediated gonadotropin regulation. *Mol Cell Endocrinol.* 2014;385:56–61.
165. Fortin J, Ongaro L, Li Y, et al. Minireview: activin signaling in gonadotropes: what does the fox say. to the SMAD? *Mol Endocrinol.* 2015;29:963–977.
166. Lewis KA, Gray PC, Blount AL, et al. Betaglycan binds inhibin and can mediate functional antagonism of activin signalling. *Nature.* 2000;404:411–414.
167. MacConell LA, Leal AM, Vale WW. The distribution of betaglycan protein and mRNA in rat brain, pituitary, and gonads: implications for a role for betaglycan in inhibin-mediated reproductive functions. *Endocrinology.* 2002;143:1066–1075.
168. Kanasaki H, Purwana IN, Mijiddorj T, Oride A, Miyazaki K. Possible involvement of PACAP and PACAP type 1 receptor in GnRH-induced FSH beta-subunit gene expression. *Regul Pept.* 2011;167:227–232.
169. Kanasaki H, Purwana IN, Miyazaki K. Possible role of PACAP and its PAC1 receptor in the differential regulation of pituitary LHbeta- and FSHbeta-subunit gene expression by pulsatile GnRH stimulation. *Biol Reprod.* 2013;88:35.
170. Purwana IN, Kanasaki H, Oride A, Mijiddorj T, Miyazaki K. Expression of the pituitary adenylate cyclase-activating polypeptide (PACAP) type 1 receptor (PAC1R) potentiates the effects of GnRH on gonadotropin subunit gene expression. *Biochem Biophys Res Commun.* 2011;410:295–300.
171. Jablonka-Shariff A, Pearl CA, Comstock A, Boime I. A carboxyl-terminal sequence in the lutropin beta subunit contributes to the sorting of lutropin to the regulated pathway. *J Biol Chem.* 2008;283:11485–11492.
172. Pearl CA, Jablonka-Shariff A, Boime I. Rerouting of a follicle-stimulating hormone analog to the regulated secretory pathway. *Endocrinology.* 2010;151:388–393.

173. Wang H, Larson M, Jablonka-Shariff A, et al. Redirecting intracellular trafficking and the secretion pattern of FSH dramatically enhances ovarian function in mice. *Proc Natl Acad Sci USA.* 2014;111:5735–5740.
174. Padmanabhan V, McFadden K, Mauger DT, Karsch FJ, Midgley Jr AR. Neuroendocrine control of follicle-stimulating hormone (FSH) secretion. I. Direct evidence for separate episodic and basal components of FSH secretion. *Endocrinology.* 1997;138:424–432.
175. Veldhuis JD, Carlson ML, Johnson ML. The pituitary gland secretes in bursts: appraising the nature of glandular secretory impulses by simultaneous multiple-parameter deconvolution of plasma hormone concentrations. *Proc Natl Acad Sci USA.* 1987;84: 7686–7690.
176. Veldhuis JD, Iranmanesh A, Johnson ML, Lizarralde G. Twenty-four-hour rhythms in plasma concentrations of adenohypophyseal hormones are generated by distinct amplitude and/or frequency modulation of underlying pituitary secretory bursts. *J Clin Endocrinol Metab.* 1990;71:1616–1623.
177. Evans WS, Sollenberger MJ, Booth Jr RA et al. Contemporary aspects of discrete peak-detection algorithms. II. The paradigm of the luteinizing hormone pulse signal in women. *Endocr Rev.* 1992;13:81–104.
178. Faria AC, Bekenstein LW, Booth Jr RA et al. Pulsatile growth hormone release in normal women during the menstrual cycle. *Clin Endocrinol.* 1992;36:591–596.
179. Filicori M, Santoro N, Merriam GR, Crowley Jr WF. Characterization of the physiological pattern of episodic gonadotropin secretion throughout the human menstrual cycle. *J Clin Endocrinol Metab.* 1986;62:1136–1144.
180. Adams JM, Taylor AE, Schoenfeld DA, Crowley Jr WF, Hall JE. The midcycle gonadotropin surge in normal women occurs in the face of an unchanging gonadotropin-releasing hormone pulse frequency. *J Clin Endocrinol Metab.* 1994;79: 858–864.
181. Hall JE, Schoenfeld DA, Martin KA, Crowley Jr WF. Hypothalamic gonadotropin-releasing hormone secretion and follicle-stimulating hormone dynamics during the luteal-follicular transition. *J Clin Endocrinol Metab.* 1992;74:600–607.
182. Wu FC, Butler GE, Kelnar CJ, Huhtaniemi I, Veldhuis JD. Ontogeny of pulsatile gonadotropin releasing hormone secretion from midchildhood, through puberty, to adulthood in the human male: a study using deconvolution analysis and an ultrasensitive immunofluorometric assay. *J Clin Endocrinol Metab.* 1996;81: 1798–1805.
183. Dunkel L, Alfthan H, Stenman UH, Tapanainen P, Perheentupa J. Pulsatile secretion of LH and FSH in prepubertal and early pubertal boys revealed by ultrasensitive time-resolved immunofluorometric assays. *Pediatr Res.* 1990;27:215–219.
184. Dunkel L, Alfthan H, Stenman UH, Perheentupa J. Gonadal control of pulsatile secretion of luteinizing hormone and follicle-stimulating hormone in prepubertal boys evaluated by ultrasensitive time-resolved immunofluorometric assays. *J Clin Endocrinol Metab.* 1990;70:107–114.
185. Urban RJ, Kaiser DL, van Cauter E, Johnson ML, Veldhuis JD. Comparative assessments of objective peak-detection algorithms. II. Studies in men. *Am J Physiol.* 1988;254:E113–E119.
186. Veldhuis JD, Evans WS, Urban RJ, Rogol AD, Johnson ML. Physiologic attributes of the luteinizing hormone pulse signal in the human. Cross-validation studies in men. *J Androl.* 1988;9:69–77.
187. Hayes FJ, Pitteloud N, DeCruz S, Crowley Jr WF, Boepple PA. Importance of inhibin B in the regulation of FSH secretion in the human male. *J Clin Endocrinol Metab.* 2001;86:5541–5546.

188. Schnorr JA, Bray MJ, Veldhuis JD. Aromatization mediates testosterone's short-term feedback restraint of 24-hour endogenously driven and acute exogenous gonadotropin-releasing hormone-stimulated luteinizing hormone and follicle-stimulating hormone secretion in young men. *J Clin Endocrinol Metab.* 2001;86:2600–2606.
189. Medina M, Ulloa-Aguirre A, Fernandez MA, Perez-Palacios G. The role of oestrogens on gonadotrophin secretion in the testicular feminization syndrome. *Acta Endocrinol.* 1980;95:314–318.
190. Liu PY, Takahashi PY, Roebuck PD, Iranmanesh A, Veldhuis JD. Aging in healthy men impairs recombinant human luteinizing hormone (LH)-stimulated testosterone secretion monitored under a two-day intravenous pulsatile LH clamp. *J Clin Endocrinol Metab.* 2005;90:5544–5550.
191. Liu PY, Takahashi PY, Roebuck PD, Iranmanesh A, Veldhuis JD. Age-specific changes in the regulation of LH-dependent testosterone secretion: assessing responsiveness to varying endogenous gonadotropin output in normal men. *Am J Physiol Regul Integr Comp Physiol.* 2005;289:R721–R728.
192. Liu PY, Iranmanesh A, Nehra AX, Keenan DM, Veldhuis JD. Mechanisms of hypoandrogenemia in healthy aging men. *Endocrinol Metab Clin North Am.* 2005;34:935–955.
193. Veldhuis JD, Keenan DM, Iranmanesh A. Mechanisms of ensemble failure of the male gonadal axis in aging. *J Endocrinol Invest.* 2005;28:8–13.
194. Veldhuis JD, Veldhuis NJ, Keenan DM, Iranmanesh A. Age diminishes the testicular steroidogenic response to repeated intravenous pulses of recombinant human LH during acute GnRH-receptor blockade in healthy men. *Am J Physiol Endocrinol Metab.* 2005;288:E775–E781.
195. Morell AG, Gregoriadis G, Scheinberg IH, Hickman J, Ashwell G. The role of sialic acid in determining the survival of glycoproteins in the circulation. *J Biol Chem.* 1971;246:1461–1467.
196. Lefort GP, Stolk JM, Nisula BC. Evidence that desialylation and uptake by hepatic receptors for galactose-terminated glycoproteins are immaterial to the metabolism of human choriogonadotropin in the rat. *Endocrinology.* 1984;115:1551–1557.
197. Blithe DL, Wehmann RE, Nisula BC. Carbohydrate composition of beta-core. *Endocrinology.* 1989;125:2267–2272.
198. Barrios-De-Tomasi J, Timossi C, Merchant H, et al. Assessment of the in vitro and in vivo biological activities of the human follicle-stimulating isohormones. *Mol Cell Endocrinol.* 2002;186:189–198.
199. Diaz-Cueto L, Barrios-de-Tomasi J, Timossi C, Mendez JP, Ulloa-Aguirre A. More in-vitro bioactive, shorter-lived human chorionic gonadotrophin charge isoforms increase at the end of the first and during the third trimesters of gestation. *Mol Hum Reprod.* 1996;2:643–650.
200. Ulloa-Aguirre A, Cravioto A, Damian-Matsumura P, Jimenez M, Zambrano E, Diaz-Sanchez V. Biological characterization of the naturally occurring analogues of intrapituitary human follicle-stimulating hormone. *Hum Reprod.* 1992;7:23–30.
201. Wide L. The regulation of metabolic clearance rate of human FSH in mice by variation of the molecular structure of the hormone. *Acta Endocrinol.* 1986;112:336–344.
202. Fiete D, Baenziger JU. Isolation of the SO4-4-GalNAcbeta1,4GlcNAcbeta1,2Manalpha-specific receptor from rat liver. *J Biol Chem.* 1997;272:14629–14637.
203. Fiete D, Srivastava V, Hindsgaul O, Baenziger JU. A hepatic reticuloendothelial cell receptor specific for SO4-4GalNAc beta 1,4GlcNAc beta 1,2Man alpha that mediates rapid clearance of lutropin. *Cell.* 1991;67:1103–1110.
204. Legardinier S, Duonor-Cerutti M, Devauchelle G, Combarnous Y, Cahoreau C. Biological activities of recombinant equine luteinizing hormone/chorionic gonadotropin (eLH/CG) expressed in Sf9 and Mimic insect cell lines. *J Mol Endocrinol.* 2005;34:47–60.

205. Duijkers IJ, Klipping C, Boerrigter PJ, Machielsen CS, De Bie JJ, Voortman G. Single dose pharmacokinetics and effects on follicular growth and serum hormones of a long-acting recombinant FSH preparation (FSH-CTP) in healthy pituitary-suppressed females. *Hum Reprod.* 2002;17:1987–1993.
206. Bouloux PM, Handelsman DJ, Jockenhovel F, et al. First human exposure to FSH-CTP in hypogonadotrophic hypogonadal males. *Hum Reprod.* 2001;16:1592–1597.
207. Perlman S, van den Hazel B, Christiansen J, et al. Glycosylation of an N-terminal extension prolongs the half-life and increases the in vivo activity of follicle stimulating hormone. *J Clin Endocrinol Metab.* 2003;88:3227–3235.
208. Weenen C, Pena JE, Pollak SV, et al. Long-acting follicle-stimulating hormone analogs containing N-linked glycosylation exhibited increased bioactivity compared with o-linked analogs in female rats. *J Clin Endocrinol Metab.* 2004;89:5204–5212.
209. Sharpless JL, Supko JG, Martin KA, Hall JE. Disappearance of endogenous luteinizing hormone is prolonged in postmenopausal women. *J Clin Endocrinol Metab.* 1999;84: 688–694.
210. Wide L. Median charge and charge heterogeneity of human pituitary FSH, LH and TSH. II. Relationship to sex and age. *Acta Endocrinol.* 1985;109:190–197.
211. Wide L. Median charge and charge heterogeneity of human pituitary FSH, LH and TSH. I. Zone electrophoresis in agarose suspension. *Acta Endocrinol.* 1985;109: 181–189.
212. Wide L, Naessen T, Sundstrom-Poromaa I, Eriksson K. Sulfonation and sialylation of gonadotropins in women during the menstrual cycle, after menopause, and with polycystic ovarian syndrome and in men. *J Clin Endocrinol Metab.* 2007;92: 4410–4417.
213. le Cotonnec JY, Loumaye E, Porchet HC, Beltrami V, Munafo A. Pharmacokinetic and pharmacodynamic interactions between recombinant human luteinizing hormone and recombinant human follicle-stimulating hormone. *Fertil Steril.* 1998;69:201–209.
214. le Cotonnec JY, Porchet HC, Beltrami V, Howles C. Comparative pharmacokinetics of two urinary human follicle stimulating hormone preparations in healthy female and male volunteers. *Hum Reprod.* 1993;8:1604–1611.
215. le Cotonnec JY, Porchet HC, Beltrami V, Khan A, Toon S, Rowland M. Clinical pharmacology of recombinant human follicle-stimulating hormone (FSH). I. Comparative pharmacokinetics with urinary human FSH. *Fertil Steril.* 1994;61: 669–678.
216. le Cotonnec JY, Porchet HC, Beltrami V, Khan A, Toon S, Rowland M. Comprehensive pharmacokinetics of urinary human follicle stimulating hormone in healthy female volunteers. *Pharm Res.* 1995;12:844–850.
217. le Cotonnec JY, Porchet HC, Beltrami V, Munafo A. Clinical pharmacology of recombinant human luteinizing hormone: Part II. Bioavailability of recombinant human luteinizing hormone assessed with an immunoassay and an in vitro bioassay. *Fertil Steril.* 1998;69:195–200.
218. le Cotonnec JY, Porchet HC, Beltrami V, Munafo A. Clinical pharmacology of recombinant human luteinizing hormone: Part I. Pharmacokinetics after intravenous administration to healthy female volunteers and comparison with urinary human luteinizing hormone. *Fertil Steril.* 1998;69:189–194.
219. Porchet HC, Le Cotonnec JY, Canali S, Zanolo G. Pharmacokinetics of recombinant human follicle stimulating hormone after intravenous, intramuscular, and subcutaneous administration in monkeys, and comparison with intravenous administration of urinary follicle stimulating hormone. *Drug Metab Dispos.* 1993;21:144–150.
220. Porchet HC, le Cotonnec JY, Loumaye E. Clinical pharmacology of recombinant human follicle-stimulating hormone. III. Pharmacokinetic-pharmacodynamic modeling after repeated subcutaneous administration. *Fertil Steril.* 1994;61:687–695.

221. Porchet HC, Le Cotonnec JY, Neuteboom B, Canali S, Zanolo G. Pharmacokinetics of recombinant human luteinizing hormone after intravenous, intramuscular, and subcutaneous administration in monkeys and comparison with intravenous administration of pituitary human luteinizing hormone. *J Clin Endocrinol Metab.* 1995;80:667–673.
222. Timossi C, Damian-Matsumura P, Dominguez-Gonzalez A, Ulloa-Aguirre A. A less acidic human follicle-stimulating hormone preparation induces tissue-type plasminogen activator enzyme activity earlier than a predominantly acidic analogue in phenobarbital-blocked pro-oestrous rats. *Mol Hum Reprod.* 1998;4:1032–1038.
223. Mannaerts BM, Rombout F, Out HJ, Coelingh Bennink H. Clinical profiling of recombinant follicle stimulating hormone (rFSH; Puregon): relationship between serum FSH and efficacy. *Hum Reprod Update.* 1996;2:153–161.
224. Smitz J, Wolfenson C, Chappel S, Ruman J. Follicle-stimulating hormone: a review of form and function in the treatment of infertility. *Reprod Sci.* 2015;23(6):706–716.
225. Rose TH, Roshammar D, Erichsen L, Grundemar L, Ottesen JT. Characterisation of population pharmacokinetics and endogenous follicle-stimulating hormone (FSH) levels after multiple dosing of a recombinant human FSH (FE 999049) in healthy women. *Drugs R D.* 2016;16(2):165–172.
226. Rose TH, Roshammar D, Erichsen L, Grundemar L, Ottesen JT. Population pharmacokinetic modelling of FE 999049, a recombinant human follicle-stimulating hormone, in healthy women after single ascending doses. *Drugs R D.* 2016;16(2):173–180.
227. Bosch E, Nyboe Andersen A, Barri P, et al. Follicular and endocrine dose responses according to anti-Mullerian hormone levels in IVF patients treated with a novel human recombinant FSH (FE 999049). *Clin Endocrinol.* 2015;83:902–912.
228. Fares FA, Suganuma N, Nishimori K, LaPolt PS, Hsueh AJ, Boime I. Design of a long-acting follitropin agonist by fusing the C-terminal sequence of the chorionic gonadotropin beta subunit to the follitropin beta subunit. *Proc Natl Acad Sci USA.* 1992;89:4304–4308.
229. Devroey P, Boostanfar R, Koper NP, Mannaerts BM, Ijzerman-Boon PC, Fauser BC. A double-blind, non-inferiority RCT comparing corifollitropin alfa and recombinant FSH during the first seven days of ovarian stimulation using a GnRH antagonist protocol. *Hum Reprod.* 2009;24:3063–3072.
230. Fauser BC, Mannaerts BM, Devroey P, Leader A, Boime I, Baird DT. Advances in recombinant DNA technology: corifollitropin alfa, a hybrid molecule with sustained follicle-stimulating activity and reduced injection frequency. *Hum Reprod Update.* 2009;15:309–321.
231. Arey BJ, Deecher DC, Shen ES, et al. Identification and characterization of a selective, nonpeptide follicle-stimulating hormone receptor antagonist. *Endocrinology.* 2002;143: 3822–3829.
232. Arey BJ, Lopez FJ. Are circulating gonadotropin isoforms naturally occurring biased agonists?. Basic and therapeutic implications. *Rev Endocr Metab Disord.* 2011;12: 275–288.
233. Nagorny P, Fasching B, Li X, Chen G, Aussedat B, Danishefsky SJ. Toward fully synthetic homogeneous beta-human follicle-stimulating hormone (beta-hFSH) with a biantennary N-linked dodecasaccharide. Synthesis of beta-hFSH with chitobiose units at the natural linkage sites. *J Am Chem Soc.* 2009;131:5792–5799.
234. Wehbi V, Decourtye J, Piketty V, Durand G, Reiter E, Maurel MC. Selective modulation of follicle-stimulating hormone signaling pathways with enhancing equine chorionic gonadotropin/antibody immune complexes. *Endocrinology.* 2010;151: 2788–2799.
235. Ulloa-Aguirre A, Crepieux P, Poupon A, Maurel MC, Reiter E. Novel pathways in gonadotropin receptor signaling and biased agonism. *Rev Endocr Metab Disord.* 2011;12:259–274.

236. Sykiotis GP, Hoang XH, Avbelj M, et al. Congenital idiopathic hypogonadotropic hypogonadism: evidence of defects in the hypothalamus, pituitary, and testes. *J Clin Endocrinol Metab.* 2010;95:3019–3027.
237. Bougneres P, Francois M, Pantalone L, et al. Effects of an early postnatal treatment of hypogonadotropic hypogonadism with a continuous subcutaneous infusion of recombinant follicle-stimulating hormone and luteinizing hormone. *J Clin Endocrinol Metab.* 2008;93:2202–2205.
238. Main KM, Schmidt IM, Skakkebaek NE. A possible role for reproductive hormones in newborn boys: progressive hypogonadism without the postnatal testosterone peak. *J Clin Endocrinol Metab.* 2000;85:4905–4907.
239. Main KM, Schmidt IM, Toppari J, Skakkebaek NE. Early postnatal treatment of hypogonadotropic hypogonadism with recombinant human FSH and LH. *Eur J Endocrinol.* 2002;146:75–79.
240. Raivio T, Toppari J, Perheentupa A, McNeilly AS, Dunkel L. Treatment of prepubertal gonadotrophin-deficient boys with recombinant human follicle-stimulating hormone. *Lancet.* 1997;350:263–264.
241. Liu PY, Baker HW, Jayadev V, Zacharin M, Conway AJ, Handelsman DJ. Induction of spermatogenesis and fertility during gonadotropin treatment of gonadotropin-deficient infertile men: predictors of fertility outcome. *J Clin Endocrinol Metab.* 2009;94:801–808.
242. Warne DW, Decosterd G, Okada H, Yano Y, Koide N, Howles CM. A combined analysis of data to identify predictive factors for spermatogenesis in men with hypogonadotropic hypogonadism treated with recombinant human follicle-stimulating hormone and human chorionic gonadotropin. *Fertil Steril.* 2009;92:594–604.
243. Pitteloud N, Hayes FJ, Boepple PA, et al. The role of prior pubertal development, biochemical markers of testicular maturation, and genetics in elucidating the phenotypic heterogeneity of idiopathic hypogonadotropic hypogonadism. *J Clin Endocrinol Metab.* 2002;87:152–160.
244. Chemes HE, Rey RA, Nistal M, et al. Physiological androgen insensitivity of the fetal, neonatal, and early infantile testis is explained by the ontogeny of the androgen receptor expression in Sertoli cells. *J Clin Endocrinol Metab.* 2008;93:4408–4412.
245. Boehm U, Bouloux PM, Dattani MT, et al. Expert consensus document: European Consensus Statement on congenital hypogonadotropic hypogonadism—pathogenesis, diagnosis and treatment. *Nat Rev Endocrinol.* 2015;11:547–564.
246. van Brakel J, Kranse R, de Muinck Keizer-Schrama SM, et al. Fertility potential in men with a history of congenital undescended testes: a long-term follow-up study. *Andrology.* 2013;1:100–108.
247. Burris AS, Rodbard HW, Winters SJ, Sherins RJ. Gonadotropin therapy in men with isolated hypogonadotropic hypogonadism: the response to human chorionic gonadotropin is predicted by initial testicular size. *J Clin Endocrinol Metab.* 1988;66:1144–1151.
248. Miyagawa Y, Tsujimura A, Matsumiya K, et al. Outcome of gonadotropin therapy for male hypogonadotropic hypogonadism at university affiliated male infertility centers: a 30-year retrospective study. *J Urol.* 2005;173:2072–2075.
249. Pitteloud N, Hayes FJ, Dwyer A, Boepple PA, Lee H, Crowley Jr WF. Predictors of outcome of long-term GnRH therapy in men with idiopathic hypogonadotropic hypogonadism. *J Clin Endocrinol Metab.* 2002;87:4128–4136.
250. Yang L, Zhang SX, Dong Q, Xiong ZB, Li X. Application of hormonal treatment in hypogonadotropic hypogonadism: more than ten years experience. *Int Urol Nephrol.* 2012;44:393–399.
251. Hoffman AR, Crowley Jr WF. Induction of puberty in men by long-term pulsatile administration of low-dose gonadotropin-releasing hormone. *N Engl J Med.* 1982;307:1237–1241.

252. Liu L, Banks SM, Barnes KM, Sherins RJ. Two-year comparison of testicular responses to pulsatile gonadotropin-releasing hormone and exogenous gonadotropins from the inception of therapy in men with isolated hypogonadotropic hypogonadism. *J Clin Endocrinol Metab.* 1988;67:1140–1145.
253. Delemarre-Van-de-Waal HA. Induction of testicular growth and spermatogenesis by pulsatile, intravenous administration of gonadotrophin-releasing hormone in patients with hypogonadotrophic hypogonadism. *Clin Endocrinol.* 1993;38:473–480.
254. Burgues S, Calderon MD. Subcutaneous self-administration of highly purified follicle stimulating hormone and human chorionic gonadotrophin for the treatment of male hypogonadotrophic hypogonadism. Spanish Collaborative Group on Male Hypogonadotropic Hypogonadism. *Hum Reprod.* 1997;12:980–986.
255. Cisternino M, Manzoni SM, Coslovich E, Autelli M. Hormonal replacement therapy with HCG and HU-FSH in thalassaemic patients affected by hypogonadotropic hypogonadism. *J Pediatr Endocrinol Metab.* 1998;11(Suppl 3):885–890.
256. Buchter D, Behre HM, Kliesch S, Nieschlag E. Pulsatile GnRH or human chorionic gonadotropin/human menopausal gonadotropin as effective treatment for men with hypogonadotropic hypogonadism: a review of 42 cases. *Eur J Endocrinol.* 1998;139:298–303.
257. Barrio R, de Luis D, Alonso M, Lamas A, Moreno JC. Induction of puberty with human chorionic gonadotropin and follicle-stimulating hormone in adolescent males with hypogonadotropic hypogonadism. *Fertil Steril.* 1999;71:244–248.
258. Bouvattier C, Tauber M, Jouret B, Chaussain JL, Rochiccioli P. Gonadotropin treatment of hypogonadotropic hypogonadal adolescents. *J Pediatr Endocrinol Metab.* 1999;12(Suppl 1):339–344.
259. Liu PY, Turner L, Rushford D, et al. Efficacy and safety of recombinant human follicle stimulating hormone (Gonal-F) with urinary human chorionic gonadotrophin for induction of spermatogenesis and fertility in gonadotrophin-deficient men. *Hum Reprod.* 1999;14:1540–1545.
260. Bouloux P, Warne DW, Loumaye E. Efficacy and safety of recombinant human follicle-stimulating hormone in men with isolated hypogonadotropic hypogonadism. *Fertil Steril.* 2002;77:270–273.
261. Bouloux PM, Nieschlag E, Burger HG, et al. Induction of spermatogenesis by recombinant follicle-stimulating hormone (puregon) in hypogonadotropic azoospermic men who failed to respond to human chorionic gonadotropin alone. *J Androl.* 2003;24: 604–611.
262. Delemarre-Van-de-Waal HA. Application of gonadotropin releasing hormone in hypogonadotropic hypogonadism. Diagnostic and therapeutic aspects. *Eur J Endocrinol.* 2004;151:U89–U94.
263. Raivio T, Wikstrom AM, Dunkel L. Treatment of gonadotropin-deficient boys with recombinant human FSH: long-term observation and outcome. *Eur J Endocrinol.* 2007;156:105–111.
264. Zacharin M, Sabin MA, Nair VV, Dabadghao P. Addition of recombinant follicle-stimulating hormone to human chorionic gonadotropin treatment in adolescents and young adults with hypogonadotropic hypogonadism promotes normal testicular growth and may promote early spermatogenesis. *Fertil Steril.* 2012;98:836–842.
265. Dwyer AA, Sykiotis GP, Hayes FJ, et al. Trial of recombinant follicle-stimulating hormone pretreatment for GnRH-induced fertility in patients with congenital hypogonadotropic hypogonadism. *J Clin Endocrinol Metab.* 2013;98:E1790–E1795.
266. Gong C, Liu Y, Qin M, Wu D, Wang X. Pulsatile GnRH is superior to hCG in therapeutic efficacy in adolescent boys with hypogonadotropic hypogonadodism. *J Clin Endocrinol Metab.* 2015;100:2793–2799.
267. Young J. Approach to the male patient with congenital hypogonadotropic hypogonadism. *J Clin Endocrinol Metab.* 2012;97:707–718.

268. Han TS, Bouloux PM. What is the optimal therapy for young males with hypogonadotropic hypogonadism? *Clin Endocrinol.* 2010;72:731–737.
269. Giltay JC, Deege M, Blankenstein RA, Kastrop PM, Wijmenga C, Lock TT. Apparent primary follicle-stimulating hormone deficiency is a rare cause of treatable male infertility. *Fertil Steril.* 2004;81:693–696.
270. Murao K, Imachi H, Muraoka T, et al. Isolated follicle-stimulating hormone (FSH) deficiency without mutation of the FSHbeta gene and successful treatment with human menopausal gonadotropin. *Fertil Steril.* 2008;90:2012.
271. Delemarre BJ. Pulsatile GnRH treatment in boys and girls with idiopathic hypogonadotrophic hypogonadism. *Hum Reprod.* 1993;8:180–183.
272. Depenbusch M, von Eckardstein S, Simoni M, Nieschlag E. Maintenance of spermatogenesis in hypogonadotropic hypogonadal men with human chorionic gonadotropin alone. *Eur J Endocrinol.* 2002;147:617–624.
273. Ley SB, Leonard JM. Male hypogonadotropic hypogonadism: factors influencing response to human chorionic gonadotropin and human menopausal gonadotropin, including prior exogenous androgens. *J Clin Endocrinol Metab.* 1985;61:746–752.
274. Raivio T, Falardeau J, Dwyer A, et al. Reversal of idiopathic hypogonadotropic hypogonadism. *N Engl J Med.* 2007;357:863–873.
275. Sidhoum VF, Chan YM, Lippincott MF, et al. Reversal and relapse of hypogonadotropic hypogonadism: resilience and fragility of the reproductive neuroendocrine system. *J Clin Endocrinol Metab.* 2014;99:861–870.
276. Pitteloud N, Boepple PA, DeCruz S, Valkenburgh SB, Crowley Jr WF, Hayes FJ. The fertile eunuch variant of idiopathic hypogonadotropic hypogonadism: spontaneous reversal associated with a homozygous mutation in the gonadotropin-releasing hormone receptor. *J Clin Endocrinol Metab.* 2001;86:2470–2475.
277. Santhakumar A, Balasubramanian R, Miller M, Quinton R. Reversal of isolated hypogonadotropic hypogonadism: long-term integrity of hypothalamo-pituitary-testicular axis in two men is dependent on intermittent androgen exposure. *Clin Endocrinol.* 2014;81:473–476.
278. King TF, Hayes FJ. Long-term outcome of idiopathic hypogonadotropic hypogonadism. *Curr Opin Endocrinol Diabetes Obes.* 2012;19:204–210.
279. Tuttelmann F, Laan M, Grigorova M, Punab M, Sober S, Gromoll J. Combined effects of the variants FSHB -211G>T and FSHR 2039A>G on male reproductive parameters. *J Clin Endocrinol Metab.* 2012;97:3639–3647.
280. Hotaling JM. Genetics of male infertility. *Urol Clin North Am.* 2014;41:1–17.
281. Gunes S, Arslan MA, Hekim GN, Asci R. The role of epigenetics in idiopathic male infertility. *J Assist Reprod Genet.* 2016;33:553–569.
282. Krausz C. Male infertility: pathogenesis and clinical diagnosis. *Best Pract Res Clin Endocrinol Metab.* 2011;25:271–285.
283. de Kretser DM. Male infertility. *Lancet.* 1997;349:787–790.
284. Hotaling JM, Patel Z. Male endocrine dysfunction. *Urol Clin North Am.* 2014;41:39–53.
285. Acosta AA, Khalifa E, Oehninger S. Pure human follicle stimulating hormone has a role in the treatment of severe male infertility by assisted reproduction: Norfolk's total experience. *Hum Reprod.* 1992;7:1067–1072.
286. Arnaldi G, Balercia G, Barbatelli G, Mantero F. Effects of long-term treatment with human pure follicle-stimulating hormone on semen parameters and sperm-cell ultrastructure in idiopathic oligoteratoasthenozoospermia. *Andrologia.* 2000;32:155–161.
287. Ashkenazi J, Bar-Hava I, Farhi J, et al. The role of purified follicle stimulating hormone therapy in the male partner before intracytoplasmic sperm injection. *Fertil Steril.* 1999;72:670–673.
288. Attia AM, Abou-Setta AM, Al-Inany HG. Gonadotrophins for idiopathic male factor subfertility. *Cochrane Database Syst Rev.* 2013;8:CD005071.

289. Baccetti B, Piomboni P, Bruni E, et al. Effect of follicle-stimulating hormone on sperm quality and pregnancy rate. *Asian J Androl.* 2004;6:133–137.
290. Baccetti B, Strehler E, Capitani S, et al. The effect of follicle stimulating hormone therapy on human sperm structure (Notulae seminologicae 11). *Hum Reprod.* 1997;12:1955–1968.
291. Bartoov B, Eltes F, Lunenfeld E, Har-Even D, Lederman H, Lunenfeld B. Sperm quality of subfertile males before and after treatment with human follicle-stimulating hormone. *Fertil Steril.* 1994;61:727–734.
292. Ben-Rafael Z, Farhi J, Feldberg D, et al. Follicle-stimulating hormone treatment for men with idiopathic oligoteratoasthenozoospermia before in vitro fertilization: the impact on sperm microstructure and fertilization potential. *Fertil Steril.* 2000;73:24–30.
293. Caprio F, De Franciscis P, Trotta C, Ianniello R, Mele D, Colacurci N. Seminal anti-Mullerian hormone levels during recombinant human follicle-stimulating hormone treatment in men with idiopathic infertility undergoing assisted reproduction cycles. *Andrology.* 2015;3:843–847.
294. Caroppo E, Niederberger C, Vizziello GM, D'Amato G. Recombinant human follicle-stimulating hormone as a pretreatment for idiopathic oligoasthenoteratozoospermic patients undergoing intracytoplasmic sperm injection. *Fertil Steril.* 2003;80:1398–1403.
295. Colacurci N, Monti MG, Fornaro F, et al. Recombinant human FSH reduces sperm DNA fragmentation in men with idiopathic oligoasthenoteratozoospermia. *J Androl.* 2012;33:588–593.
296. Ding YM, Zhang XJ, Li JP, et al. Treatment of idiopathic oligozoospermia with recombinant human follicle-stimulating hormone: a prospective, randomized, double-blind, placebo-controlled clinical study in Chinese population. *Clin Endocrinol.* 2015;83:866–871.
297. Dirnfeld M, Katz G, Calderon I, Abramovici H, Bider D. Pure follicle-stimulating hormone as an adjuvant therapy for selected cases in male infertility during in-vitro fertilization is beneficial. *Eur J Obstet Gynecol Reprod Biol.* 2000;93:105–108.
298. Efesoy O, Cayan S, Akbay E. The efficacy of recombinant human follicle-stimulating hormone in the treatment of various types of male-factor infertility at a single university hospital. *J Androl.* 2009;30:679–684.
299. Farrag A, Sagnella F, Pappalardo S, et al. The use of r-hFSH in treatment of idiopathic male factor infertility before ICSI. *Eur Rev Med Pharmacol Sci.* 2015;19:2162–2167.
300. Fernandez-Arjona M, Diaz J, Cortes I, Gonzalez J, Rodriguez JM, Alvarez E. Relationship between gonadotrophin secretion, inhibin B and spermatogenesis in oligozoospermic men treated with highly purified urinary follicle-stimulating hormone (uFSH-HP): a preliminary report. *Eur J Obstet Gynecol Reprod Biol.* 2003;107:47–51.
301. Foresta C, Bettella A, Ferlin A, Garolla A, Rossato M. Evidence for a stimulatory role of follicle-stimulating hormone on the spermatogonial population in adult males. *Fertil Steril.* 1998;69:636–642.
302. Foresta C, Bettella A, Garolla A, Ambrosini G, Ferlin A. Treatment of male idiopathic infertility with recombinant human follicle-stimulating hormone: a prospective, controlled, randomized clinical study. *Fertil Steril.* 2005;84:654–661.
303. Foresta C, Bettella A, Merico M, Garolla A, Ferlin A, Rossato M. Use of recombinant human follicle-stimulating hormone in the treatment of male factor infertility. *Fertil Steril.* 2002;77:238–244.
304. Foresta C, Bettella A, Merico M, et al. FSH in the treatment of oligozoospermia. *Mol Cell Endocrinol.* 2000;161:89–97.
305. Iacono F, Barra S, Montano L, Lotti T. Value of high-dose pure FSH in the treatment of idiopathic male infertility. *J Urol.* 1996;102:81–84.
306. Kamischke A, Behre HM, Bergmann M, Simoni M, Schafer T, Nieschlag E. Recombinant human follicle stimulating hormone for treatment of male idiopathic

infertility: a randomized, double-blind, placebo-controlled, clinical trial. *Hum Reprod.* 1998;13:596–603.
307. Knuth UA, Honigl W, Bals-Pratsch M, Schleicher G, Nieschlag E. Treatment of severe oligospermia with human chorionic gonadotropin/human menopausal gonadotropin: a placebo-controlled, double blind trial. *J Clin Endocrinol Metab.* 1987;65:1081–1087.
308. Palomba S, Falbo A, Espinola S, et al. Effects of highly purified follicle-stimulating hormone on sperm DNA damage in men with male idiopathic subfertility: a pilot study. *J Endocrinol Invest.* 2011;34:747–752.
309. Paradisi R, Busacchi P, Seracchioli R, Porcu E, Venturoli S. Effects of high doses of recombinant human follicle-stimulating hormone in the treatment of male factor infertility: results of a pilot study. *Fertil Steril.* 2006;86:728–731.
310. Piomboni P, Serafini F, Gambera L, et al. Sperm aneuploidies after human recombinant follicle stimulating hormone therapy in infertile males. *Reprod Biomed Online.* 2009;18:622–629.
311. Santi D, Granata AR, Simoni M. FSH treatment of male idiopathic infertility improves pregnancy rate: a meta-analysis. *Endocr Connect.* 2015;4:R46–R58.
312. Strehler E, Sterzik K, De Santo M, et al. The effect of follicle-stimulating hormone therapy on sperm quality: an ultrastructural mathematical evaluation. *J Androl.* 1997;18:439–447.
313. Valenti D, La Vignera S, Condorelli RA, et al. Follicle-stimulating hormone treatment in normogonadotropic infertile men. *Nat Rev Urol.* 2013;10:55–62.
314. Jensen TK, Andersson AM, Hjollund NH, et al. Inhibin B as a serum marker of spermatogenesis: correlation to differences in sperm concentration and follicle-stimulating hormone levels. A study of 349 Danish men. *J Clin Endocrinol Metab.* 1997;82:4059–4063.
315. Mahmoud AM, Comhaire FH, Depuydt CE. The clinical and biologic significance of serum inhibins in subfertile men. *Reprod Toxicol.* 1998;12:591–599.
316. Uhler ML, Zinaman MJ, Brown CC, Clegg ED. Relationship between sperm characteristics and hormonal parameters in normal couples. *Fertil Steril.* 2003;79 (Suppl 3):1535–1542.
317. Meeker JD, Godfrey-Bailey L, Hauser R. Relationships between serum hormone levels and semen quality among men from an infertility clinic. *J Androl.* 2007;28:397–406.
318. Andersson AM, Petersen JH, Jorgensen N, Jensen TK, Skakkebaek NE. Serum inhibin B and follicle-stimulating hormone levels as tools in the evaluation of infertile men: significance of adequate reference values from proven fertile men. *J Clin Endocrinol Metab.* 2004;89:2873–2879.
319. Wang H, Butnev V, Bousfield GR, Kumar TR. A human FSHB transgene encoding the double N-glycosylation mutant (Asn(7Delta) Asn(24Delta)) FSHbeta subunit fails to rescue Fshb null mice. *Mol Cell Endocrinol.* 2016;426:113–124.
320. Dolci S, Jannini EA, Rossi P. Gonadotropin and testicular function: paracrine control of spermatogonia self renewal, proliferation and differentiation. *Curr Trends Clin Embriol.* 2015;2:66–72.
321. Pescovitz OH, Srivastava CH, Breyer PR, Monts BA. Paracrine control of spermatogenesis. *Trends Endocrinol Metab.* 1994;5:126–131.
322. Rossi P, Dolci S. Paracrine mechanisms involved in the control of early stages of mammalian spermatogenesis. *Front Endocrinol.* 2013;4:181.
323. Huhtaniemi IT, Themmen AP. Mutations in human gonadotropin and gonadotropin-receptor genes. *Endocrine.* 2005;26:207–217.
324. Laan M, Grigorova M, Huhtaniemi IT. Pharmacogenetics of follicle-stimulating hormone action. *Curr Opin Endocrinol Diabetes Obes.* 2012;19:220–227.
325. Behre HM, Greb RR, Mempel A, et al. Significance of a common single nucleotide polymorphism in exon 10 of the follicle-stimulating hormone (FSH) receptor gene for

the ovarian response to FSH: a pharmacogenetic approach to controlled ovarian hyperstimulation. *Pharmacogenet Genomics.* 2005;15:451–456.

326. Greb RR, Grieshaber K, Gromoll J, et al. A common single nucleotide polymorphism in exon 10 of the human follicle stimulating hormone receptor is a major determinant of length and hormonal dynamics of the menstrual cycle. *J Clin Endocrinol Metab.* 2005;90:4866–4872.
327. Hoogendoorn B, Coleman SL, Guy CA, et al. Functional analysis of human promoter polymorphisms. *Hum Mol Genet.* 2003;12:2249–2254.
328. Tuttelmann F, Terwort N, Laan M, Gromoll J. FSHB promotor polymorphism influences male reproductive parameters. *J Reproduktionsmed Endokrinol.* 2011;8:305–345.
329. Hagen CP, Aksglaede L, Sorensen K, et al. FSHB-211 and FSHR 2039 are associated with serum levels of follicle-stimulating hormone and antimullerian hormone in healthy girls: a longitudinal cohort study. *Fertil Steril.* 2013;100:1089–1095.
330. Grigorova M, Punab M, Ausmees K, Laan M. FSHB promoter polymorphism within evolutionary conserved element is associated with serum FSH level in men. *Hum Reprod.* 2008;23:2160–2166.
331. Grigorova M, Punab M, Poolamets O, et al. Increased Prevalance of the -211 T allele of follicle stimulating hormone (FSH) beta subunit promoter polymorphism and lower serum FSH in infertile men. *J Clin Endocrinol Metab.* 2010;95:100–108.
332. Grigorova M, Punab M, Zilaitiene B, et al. Genetically determined dosage of follicle-stimulating hormone (FSH) affects male reproductive parameters. *J Clin Endocrinol Metab.* 2011;96:E1534–E1541.
333. Ferlin A, Vinanzi C, Selice R, Garolla A, Frigo AC, Foresta C. Toward a pharmacogenetic approach to male infertility: polymorphism of follicle-stimulating hormone beta-subunit promoter. *Fertil Steril.* 2011;96:1344–1349.
334. Casarini L, Moriondo V, Marino M, et al. FSHR polymorphism p.N680S mediates different responses to FSH in vitro. *Mol Cell Endocrinol.* 2014;393:83–91.
335. Tranchant T, Durand G, Piketty V, et al. N680S SNP of the human FSH receptor impacts on basal FSH and estradiol level in women and modifies PKA nuclear translocation and CREB-dependent gene transcription in vitro. *Hum Reprod.* 2012;27(Suppl 1):i45–i46.
336. Grigorova M, Punab M, Poolamets O, et al. Study in 1790 Baltic men: FSHR Asn680Ser polymorphism affects total testes volume. *Andrology.* 2013;1:293–300.
337. Selice R, Garolla A, Pengo M, Caretta N, Ferlin A, Foresta C. The response to FSH treatment in oligozoospermic men depends on FSH receptor gene polymorphisms. *Int J Androl.* 2011;34:306–312.
338. Olsson H, Sandstrom R, Grundemar L. Different pharmacokinetic and pharmacodynamic properties of recombinant follicle-stimulating hormone (rFSH) derived from a human cell line compared with rFSH from a non-human cell line. *J Clin Pharmacol.* 2014;54:1299–1307.
339. Arey BJ. Allosteric modulators of glycoprotein hormone receptors: discovery and therapeutic potential. *Endocrine.* 2008;34:1–10.
340. Arey BJ, Yanofsky SD, Claudia Perez M, et al. Differing pharmacological activities of thiazolidinone analogs at the FSH receptor. *Biochem Biophys Res Commun.* 2008;368: 723–728.
341. Maclean D, Holden F, Davis AM, et al. Agonists of the follicle stimulating hormone receptor from an encoded thiazolidinone library. *J Comb Chem.* 2004;6:196–206.
342. Yanofsky SD, Shen ES, Holden F, et al. Allosteric activation of the follicle-stimulating hormone (FSH) receptor by selective, nonpeptide agonists. *J Biol Chem.* 2006;281: 13226–13233.

CHAPTER FIVE

Beyond Reproduction: Pituitary Hormone Actions on Bone

T. Yuen*, L. Sun*, P. Liu*, H.C. Blair†, M. New*, A. Zallone‡, M. Zaidi*,[1]

* The Mount Sinai Bone Program, Department of Medicine, and Department of Pediatrics, Icahn School of Medicine at Mount Sinai, New York, NY, United States

† Departments of Pathology and of Cell Biology, University of Pittsburgh School of Medicine and the Pittsburgh VA Medical Center, Pittsburgh, PA, United States

‡ Department of Histology, University of Bari, Bari, Italy

[1] Corresponding author. E-mail address: mone.zaidi@mountsinai.org

Contents

Abstract

The long-held belief that pituitary hormones act solely on master targets was first questioned when we documented G protein–coupled receptors for thyroid-stimulating hormone, follicle-stimulating hormone, adrenocorticotrophic hormone, oxytocin, and vasopressin on bone cells. These evolutionarily conserved hormones and their receptors are known to have primitive roles, and exist in invertebrate species as far down as coelenterates. It is not surprising therefore that each such hormone has multiple hitherto unrecognized functions in mammalian integrative physiology, and hence, becomes a potential target for therapeutic intervention. Here we discuss the skeletal actions of pituitary hormones.

Traditionally, each anterior and posterior pituitary glycoprotein hormone has been ascribed a singular function in physiology. It has nonetheless become clear over the past decade that these hormones and their receptors have more ubiquitous functions. Organs, such as the skeleton, are regulated

Progress in Molecular Biology and Translational Science, Volume 143
ISSN 1877-1173
http://dx.doi.org/10.1016/bs.pmbts.2016.08.004

by and respond to pituitary hormones, particularly when circulating levels are perturbed in disease. Certain pituitary hormones are also expressed in bone cells, underscoring paracrine regulation. Here, we will discuss broadly the evolution of this subfamily of glycoprotein hormones and their receptors, and focus on the extragonadal, skeletal action of follicle-stimulating hormone (FSH).

1. EVOLUTIONARY CONSIDERATIONS

FSH and luteinizing hormone (LH), along with chorionic gonadotropin (hCG) and thyroid-stimulating hormone (TSH), belong to the subfamily of heterodimeric protein ligands that share a common α-subunit, with specificity depending on their distinct β-subunits. These hormones have primitive functions. In coelenterates, a TSHR family gene is identifiable, widely expressed, and shows the intron–exon structure found in mammals.[1] In lower vertebrates, such as in bony fish, there is abundant TSHR expression in the thyroid, but the receptor is also detectable in ovaries, heart, muscle, and brain.[2] Likewise, the gonadal expression of LHR and FSHR is well documented in fish. In fact, multiple differently processed forms of the FSHR occur in fish[3]; this may reflect isoforms with differing functions. Further, in fish the FSHR binds both FSH and LH, whereas the LHR recognizes only LH.[4] While high-level FSHR expression is restricted to gonads, low-level expression is seen in spleen,[5] quite similarly to findings in human cells.

Others and we have recently reported the production of a TSHβ variant by bone marrow macrophages[6,7]; this splice variant activates the TSHR, and may do so locally in the bone. Lymphocytes also express TSH,[8,9] but such production is unlikely to affect circulating levels. There is no evidence, however, for bone or marrow cell production of FSH, although coproduction of TSHβ and FSH is noted in CD11β cells from the mouse thyroid.[10]

Overall, therefore, the presence of G protein–coupled receptors (GPCRs) in tissues other than traditional endocrine targets, such as the skeleton, and in cases, coexistence of their ligands, comes as no surprise. What does come as a surprise is that the skeleton appears to be more sensitive to GPCR stimulation than the primary target organs, at least in mouse genetic and limited human studies. Thus, FSH, TSH, adrenocorticotrophic hormone (ACTH), growth hormone (GH), prolactin (PRL), oxytocin (OTX), and vasopressin all affect bone, and in mice, the haploinsufficiency of either the ligand and/or receptor yields a skeletal phenotype with the primary target organ remaining unaffected. The function of pituitary glycoprotein GPCRs in skeletal control therefore appears evolutionarily more distant than effects on the primary endocrine tissues.[11]

2. EPIDEMIOLOGY SUGGESTS AN FSH–BONE CONNECTION

Correlations between bone mineral density (BMD) declines and serum FSH levels have been documented extensively. The Study of Women's Health Across The Nations (SWAN), a longitudinal cohort of 2375 perimenopausal women, showed a strong correlation between serum FSH levels and markers of bone resorption, as well as an equally strong correlation between changes in FSH levels over 4 years and decrements in BMD.[12] Analyses of data from Chinese women showed similar trends: a significant association between bone loss and high-serum FSH.[13,14] In a group of southern Chinese women aged between 45 and 55 years, those in the highest quartile of serum FSH lost bone at a 1.3- to 2.3-fold higher rate than those in the lowest quartile.[15] Furthermore, analysis of a National Health and Nutrition Examination Survey (NHANES) III cohort of women between the ages of 42 and 60 years showed a strong correlation between serum FSH and femoral neck BMD.[16] Likewise, a cross-sectional analysis of 92 postmenopausal women found that serum osteocalcin and C-telopeptide levels were both positively correlated with FSH, but not with estradiol.[17]

The BONTURNO Study group showed that women with serum FSH levels of >30 IU/mL had a significantly higher bone turnover markers than age-matched women, despite having normal menses.[18] In contrast, Gourlay et al. fail to show a strong relationship between bone mass and FSH or indeed estrogen.[19] Interestingly, however, the same authors show an independent correlation between FSH and lean mass.[20] This latter association makes biological sense inasmuch as FSHRs are present on mesenchymal stem cells,[21] known to have the propensity for adipocyte and myocyte differentiation. Notwithstanding whether there is a cause–effect relationship between rising FSH and BMD changes, there is clear evidence favoring the use of FSH as a serum marker for identifying "fast bone losers" during the early phases of the menopausal transition, most notably the late perimenopause.[22]

3. FSH DIRECTLY AFFECTS THE SKELETON

Several studies have confirmed direct effects of FSH on the skeleton in rodents and humans. For example, amenorrheic women with a higher mean serum FSH (~35 IU/L) have greater bone loss than those with lower levels (~8 IU/L) in the face of near-equal estrogen levels.[23] Patients with functional hypothalamic amenorrhea, in whom both FSH and estrogen were low, show slight-to-moderate skeletal defects.[24] Women harboring an activating FSHR-N680S polymorphism, rs6166, have lower bone mass and high resorption markers[25]; this attests to a role for FSHRs in human physiology. The exogenous administration of FSH to rats augments ovariectomy-induced bone loss, and a FSH antagonist reduces bone loss after ovariectomy or FSH injection.[26,27]

It has been difficult to tease out the action of FSH from that of estrogen in vivo as FSH releases estrogen and the actions of FSH and estrogen on the osteoclast are opposed. The injection of FSH into mice with intact ovaries,[28] or its transgenic overexpression,[29] even in *hpg* mice, is unlikely to reveal proresorptive actions of FSH. This is because direct effects of FSH on the osteoclast will invariably be masked by the antiresorptive and anabolic actions of the ovarian estrogen so released in response to FSH.

Clinically, there is evidence that women with low FSH levels undergo less bone loss than women with higher FSH levels, estrogen levels being nearly equal,[23] and importantly, that the effectiveness of estrogen therapy is related to the degree of FSH suppression.[30] With that said, patients with pituitary hypogonadism do lose bone. Luperide treatment, and hence the lowering of FSH, has not been shown to prevent hypogonadal hyperresorption.[31] While this proves that low estrogen is a *cause of* acute hypogonadal bone loss, it does not exclude a role for FSH in human skeletal homeostasis.[31] Rather than blocking FSH in acute hypogonadism, where the effect of low estrogen is likely to be overwhelming, FSH inhibition during the late perimenopause, particularly when estrogen levels are normal and FSH is high, could potentially be of therapeutic significance.

4. FSH SIGNALS THROUGH FSHRS ON BONE CELLS

We discovered that FSH directly stimulates bone resorption by osteoclasts.[21,32] FSH increases osteoclast formation, function, and survival through a distinct FSHR isoform.[21,33–35] Two groups have, however, failed to identify FSHRs on osteoclasts, having likely used primers targeted to the ovarian isoform.[28,29] We very consistently find FSHR in human $CD14^+$ cells and osteoclasts using nested primers and sequencing to verify the specificity of the reaction, and amplifying regions that contain an intron to avoid the pitfall of genomic DNA contamination.[33]

Wu et al. show that the osteoclastogenic response to FSH was abolished in mice lacking ITAM adapter signaling molecules.[35] This suggests an interaction between FSH and immune receptor complexes, although the significance of such an interaction remains unclear. In a separate study, FSHR activation was shown to enhance RANK receptor expression.[36] In addition, FSH indirectly stimulates osteoclast formation by releasing osteoclastogenic cytokines, namely IL-1β, TNF-α, and IL-6 in proportion to the surface expression of FSHRs.[32,37] In a study of 36 women between the ages of 20 and 50, serum FSH concentrations correlated with circulating cytokine concentrations.[37,38]

5. FSH ANTIBODY RESCUES HYPOGONADAL BONE LOSS

We have recently developed a polyclonal antibody against the 13-amino acid long, conserved peptide sequence of the computationally defined FSHR-binding domain of FSHβ subunit.[39,40] We found that this antibody not only attenuated osteoclastogenesis induced by FSH in vitro, but also prevented the loss of bone postovariectomy.[39,40] Most importantly, we noted through a detailed histomorphometric analysis that while bone resorption was reduced, bone formation was surprisingly elevated significantly.[39] This decoupling likely underscored the osteoprotective action of the antibody. Further studies documented FSHRs on mesenchymal stem cells. Additionally, bone marrow stromal cells from $Fshr^{-/-}$ mice showed significantly greater colony-forming (Cfu-f) potential, suggesting a profound increase in osteoblastogenesis in the absence of the FSHR. Importantly, antibody injections did not affect serum estrogen levels in wild type mice.[39] Collectively, our studies provide clear evidence that the lowering FSH in a hypogonadal state prevents bone loss.

6. OXYTOCIN IS ANABOLIC TO BONE

OXT is a nonapeptide synthesized in the hypothalamus, stored in and released into circulation via the posterior pituitary. Its primary function is to mediate the milk ejection reflex in nursing mammals. It also stimulates uterine contraction during parturition. However, OXT is not a requirement for this function. Thus, *Oxt* null mice can deliver normally, but are unable to nurse. Subcutaneous OXT injection completely rescues the milk ejection phenotype, attesting to this being a peripheral, as opposed to a central action.[41] Central actions of OXT include the regulation of social behavior, including sexual and maternal behavior, affiliation, social memory, as well as penile erection and ejaculation.[42–45] It also controls food, predominantly carbohydrate, intake centrally.[46] Thus, the social amnesia, aggressive behavior, and overfeeding observed in *Oxt*$^{-/-}$ and *Oxtr*$^{-/-}$ mice are reversed on intracerebroventricular OXT injection.[47]

OXT acts via a GPCR, abundantly present on osteoblasts,[48] osteoclasts, and their precursors.[49] In line with the ubiquitous distribution of OXTRs, cells of bone marrow also synthesize OXT, suggesting the existence of autocrine and paracrine interactions.[50] In vitro, OXT stimulates osteoblast differentiation and bone formation. Thus, *Oxt*$^{-/-}$ and *Oxtr*$^{-/-}$ mice, including the haploinsufficient heterozygotes with normal lactation, display severe osteoporosis due to a bone-forming defect.[51] This not only indicates that the osteoblast is the target for OXT, but also that bone is more sensitive to OXT than the breast, hitherto considered its primary target. Once again, these findings emphasize a relatively primitive pituitary–bone axis. Effects of OXT on bone resorption in vivo appear minimal, as OXT stimulates osteoclastogenesis, but inhibits the activity of mature osteoclasts, with a net zero effect on resorption.

In vivo gain-of-function studies document a direct effect of OXT on bone. Intraperitoneal OXT injections result in an increased BMD and ex vivo osteoblast formation.[51] In contrast, short-term intracerebroventricular OXT does not affect bone turnover markers. OXT injections in wild-type rats alter the RANKL/Osteoprotegerin (OPG) ratio in favor on bone formation, again attesting to an anabolic action.[52]

7. OXYTOCIN AND SKELETAL REGULATION DURING REPRODUCTION

Although unproven, OXT may have a critical role in bone anabolism during pregnancy and lactation. Both are characterized by excessive bone resorption in favor of fetal and postpartum bone growth, respectively.[53] This bone loss is, however, completely reversed upon weaning by a yet unidentified mechanism.[54] OXT peaks in blood during late pregnancy and lactation, and while its proosteoclastogeneic action may contribute to intergenerational calcium transfer, its anabolic action could enable the restoration of the maternal skeleton. That *Oxt*$^{-/-}$ pups show hypomineralized s keletons, and *Oxt*$^{-/-}$ moms display reduced bone formation markers are suggestive of such actions. The question whether estrogen, via its positive regulation of osteoblastic OXT production, can synergize this action through a local feed-forward loop, remains to be determined. These studies nonetheless pave the way for a greater understanding of pregnancy and lactation-associated osteoporosis, as well as new potential therapeutic options.

Finally, the highly homologous arginine–vasopressin (AVP) and OXT can share their receptors in the regulation of bone formation by osteoblasts. AVPR1A and OXTR have opposing effects on bone mass. Notably, the deletion of *Oxtr* in *Oxt*$^{-/-}$: *Avpr1a*$^{-/-}$ double-mutant mice reversed the high-bone mass phenotype in *Avpr1a*$^{-/-}$ mice.[55] While AVP-stimulated gene expression is inhibited when the OXTR is deleted in *Avpr1a*$^{-/-}$ cells, OXTR is dispensable for AVP action in inhibiting osteoblastogenesis and gene expression.[55] Furthermore, OXT does not interact with AVPRs in vivo in a model of lactation-induced bone loss in which OXT levels are high.[55]

8. PROLACTIN AND BONE

PRL, a peptide hormone secreted by the anterior pituitary, acts to induce and maintain lactation, as well as to suppress folliculogenesis and libido. During pregnancy, it increases the calcium bioavailability for milk production and fetal skeletalogenesis by promoting intestinal calcium absorption and skeletal mobilization.[56] Accelerated bone turnover and bone loss is noted in hyperprolactinemic adults.[57] Antagonism of PRL by bromocriptine, a dopamine agonist, reverses the bone loss.[58]

This osteoclastic action of PRL is traditionally thought to arise from the accompanying hypoestrogenemia.[59] However, it has been shown that osteoblasts express PRLRs,[60] suggesting a direct interaction between PRL and the osteoblast. In fact, the pattern of bone loss is distinct in PRL-exposed and ovariectomized rats.[61] Ex vivo, PRL decreases osteoblast differentiation markers,[61] in part, through the PI3K signaling pathway.[62] PRL injected into adult mice accelerates bone resorption,[61] notably by increasing the RANK/OPG ratio.[61] Osteoclasts themselves do not possess PRLRs.[60] In contrast, in infant rats, PRL causes net bone gain,[63] increased osteocalcin expression. Likewise, in human fetal osteoblast cells, PRL decreases the RANKL/OPG ratio.[62] It appears therefore that the net effect of PLR on bone depends on the stage of development. In the fetus, it promotes bone growth and mineralization, while accelerating bone resorption in the mother to make nutrients available.

REFERENCES

1. Vibede N, Hauser F, Williamson M, Grimmelikhuijzen CJ. Genomic organization of a receptor from sea anemones, structurally and evolutionarily related to glycoprotein hormone receptors from mammals. *Biochem Biophys Res Commun.* 1998;252:497–501.
2. Kumar RS, Ijiri S, Kight K, Swanson P, Dittman A, Alok D, Zohar Y, Trant JM. Cloning and functional expression of a thyrotropin receptor from the gonads of a vertebrate (bony fish): potential thyroid-independent role for thyrotropin in reproduction. *Mol Cell Endocrinol.* 2000;167:1–9.
3. Kobayashi T, Andersen O. The gonadotropin receptors FSH-R and LH-R of Atlantic halibut (*Hippoglossushippoglossus*), 1: isolation of multiple transcripts encoding full-length and truncated variants of FSH-R. *Gen Comp Endocrinol.* 2008;156:584–594.
4. Bogerd J, Granneman JC, Schulz RW, Vischer HF. Fish FSH receptors bind LH: how to make the human FSH receptor to be more fishy? *Gen Comp Endocrinol.* 2005;142: 34–43.
5. Kumar RS, Ijiri S, Trant JM. Molecular biology of the channel catfish gonadotropin receptors: 2. Complementary DNA cloning, functional expression, and seasonal gene expression of the follicle-stimulating hormone receptor. *Biol Reprod.* 2001;65:710–717.
6. Vincent BH, Montufar-Solis D, Teng BB, Amendt BA, Schaefer J, Klein JR. Bone marrow cells produce a novel TSHbeta splice variant that is upregulated in the thyroid following systemic virus infection. *Genes Immun.* 2009;10:18–26.
7. Baliram R, Chow A, Huber AK, Collier L, Ali MR, Morshed SA, Latif R, Teixeira A, Merad M, Liu L, Sun L, Blair HC, Zaidi M, Davies TF. Thyroid and bone: macrophage-derived TSH-beta splice variant increases murine osteoblastogenesis. *Endocrinology.* 2013;154:4919–4926.
8. Smith EM, Phan M, Kruger TE, Coppenhaver DH, Blalock JE. Human lymphocyte production of immunoreactive thyrotropin. *Proc Natl Acad Sci USA.* 1983;80:6010–6013.
9. Harbour DV, Kruger TE, Coppenhaver D, Smith EM, Meyer 3rd WJ. Differential expression and regulation of thyrotropin (TSH) in T cell lines. *Mol Cell Endocrinol.* 1989;64:229–241.
10. Klein JR, Wang HC. Characterization of a novel set of resident intrathyroidal bone marrow-derived hematopoietic cells: potential for immune-endocrine interactions in thyroid homeostasis. *J Exp Biol.* 2004;207:55–65.
11. Blair HC, Robinson LJ, Sun L, Isales C, Davies TF, Zaidi M. Skeletal receptors for steroid-family regulating glycoprotein hormones: a multilevel, integrated physiological control system. *Ann NY Acad Sci.* 2011;1240:26–31.
12. Sowers MR, Greendale GA, Bondarenko I, Finkelstein JS, Cauley JA, Neer RM, Ettinger B. Endogenous hormones and bone turnover markers in pre- and perimenopausal women: SWAN. *Osteoporos Int.* 2003;14:191–197.
13. Xu ZR, Wang AH, Wu XP, Zhang H, Sheng ZF, Wu XY, Xie H, Luo XH, Liao EY. Relationship of age-related concentrations of serum FSH and LH with bone mineral density, prevalence of osteoporosis in native Chinese women. *Clin Chim Acta.* 2009;400: 8–13.

14. Wu XY, Wu XP, Xie H, Zhang H, Peng YQ, Yuan LQ, Su X, Luo XH, Liao EY. Age-related changes in biochemical markers of bone turnover and gonadotropin levels and their relationship among Chinese adult women. *Osteoporos Int.* 2010;21:275–285.
15. Cheung E, Tsang S, Bow C, Soong C, Yeung S, Loong C, Cheung CL, Kan A, Lo S, Tam S, Tang G, Kung A. Bone loss during menopausal transition among southern Chinese women. *Maturitas.* 2011;69:50–56.
16. Gallagher CM, Moonga BS, Kovach JS. Cadmium, follicle-stimulating hormone, and effects on bone in women age 42-60 years, NHANES III. *Environ Res.* 2010;110: 105–111.
17. Garcia-Martin A, Reyes-Garcia R, Garcia-Castro JM, Rozas-Moreno P, Escobar-Jimenez F, Munoz-Torres M. Role of serum FSH measurement on bone resorption in postmenopausal women. *Endocrine.* 2012;41:302–308.
18. Adami S, Bianchi G, Brandi ML, Giannini S, Ortolani S, DiMunno O, Frediani B, Rossini M. group Bs.Determinants of bone turnover markers in healthy premenopausal women. *Calcif Tissue Int.* 2008;82:341–347.
19. Gourlay ML, Preisser JS, Hammett-Stabler CA, Renner JB, Rubin J. Follicle-stimulating hormone and bioavailable estradiol are less important than weight and race in determining bone density in younger postmenopausal women. *Osteoporos Int.* 2011;22: 2699–2708.
20. Gourlay ML, Specker BL, Li C, Hammett-Stabler CA, Renner JB, Rubin JE. Follicle-stimulating hormone is independently associated with lean mass but not BMD in younger postmenopausal women. *Bone.* 2012;50:311–316.
21. Sun L, Peng Y, Sharrow AC, Iqbal J, Zhang Z, Papachristou DJ, Zaidi S, Zhu LL, Yaroslavskiy BB, Zhou H, Zallone A, Sairam MR, Kumar TR, Bo W, Braun J, Cardoso-Landa L, Schaffler MB, Moonga BS, Blair HC, Zaidi M. FSH directly regulates bone mass. *Cell.* 2006;125:247–260.
22. Zaidi M, Turner CH, Canalis E, Pacifici R, Sun L, Iqbal J, Guo XE, Silverman S, Epstein S, Rosen CJ. Bone loss or lost bone: rationale and recommendations for the diagnosis and treatment of early postmenopausal bone loss. *Curr Osteoporos Rep.* 2009;7:118–126.
23. Devleta B, Adem B, Senada S. Hypergonadotropic amenorrhea and bone density: new approach to an old problem. *J Bone Miner Metab.* 2004;22:360–364.
24. Podfigurna-Stopa A, Pludowski P, Jaworski M, Lorenc R, Genazzani AR, Meczekalski B. Skeletal status and body composition in young women with functional hypothalamic amenorrhea. *Gynecol Endocrinol.* 2012;28:299–304.
25. Rendina D, Gianfrancesco F, De Filippo G, Merlotti D, Esposito T, Mingione A, Nuti R, Strazzullo P, Mossetti G, Gennari L. FSHR gene polymorphisms influence bone mineral density and bone turnover in postmenopausal women. *Eur J Endocrinol.* 2010;163: 165–172.
26. Liu S, Cheng Y, Fan M, Chen D, Bian Z. FSH aggravates periodontitis-related bone loss in ovariectomized rats. *J Dent Res.* 2010;89:366–371.
27. Liu S, Cheng Y, Xu W, Bian Z. Protective effects of follicle-stimulating hormone inhibitor on alveolar bone loss resulting from experimental periapical lesions in ovariectomized rats. *J Endod.* 2010;36:658–663.
28. Ritter V, Thuering B, Saint Mezard P, Luong-Nguyen NH, Seltenmeyer Y, Junker U, Fournier B, Susa M, Morvan F. Follicle-stimulating hormone does not impact male bone mass in vivo or human male osteoclasts in vitro. *Calcif Tissue Int.* 2008;82:383–391.
29. Allan CM, Kalak R, Dunstan CR, McTavish KJ, Zhou H, Handelsman DJ, Seibel MJ. Follicle-stimulating hormone increases bone mass in female mice. *Proc Natl Acad Sci USA.* 2010;107:22629–22634.
30. Kawai H, Furuhashi M, Suganuma N. Serum follicle-stimulating hormone level is a predictor of bone mineral density in patients with hormone replacement therapy. *Arch Gynecol Obstet.* 2004;269:192–195.

31. Drake MT, McCready LK, Hoey KA, Atkinson EJ, Khosla S. Effects of suppression of follicle-stimulating hormone secretion on bone resorption markers in postmenopausal women. *J Clin Endocrinol Metab.* 2010;95:5063–5068.
32. Iqbal J, Sun L, Kumar TR, Blair HC, Zaidi M. Follicle-stimulating hormone stimulates TNF production from immune cells to enhance osteoblast and osteoclast formation. *Proc Natl Acad Sci USA.* 2006;103:14925–14930.
33. Robinson LJ, Tourkova I, Wang Y, Sharrow AC, Landau MS, Yaroslavskiy BB, Sun L, Zaidi M, Blair HC. FSH-receptor isoforms and FSH-dependent gene transcription in human monocytes and osteoclasts. *Biochem Biophys Res Commun.* 2010;394:12–17.
34. Sun L, Zhang Z, Zhu LL, Peng Y, Liu X, Li J, Agrawal M, Robinson LJ, Iqbal J, Blair HC, Zaidi M. Further evidence for direct pro-resorptive actions of FSH. *Biochem Biophys Res Commun.* 2010;394:6–11.
35. Wu Y, Torchia J, Yao W, Lane NE, Lanier LL, Nakamura MC, Humphrey MB. Bone microenvironment specific roles of ITAM adapter signaling during bone remodeling induced by acute estrogen-deficiency. *PLoS One.* 2007;2:e586.
36. Cannon JG, Kraj B, Sloan G. Follicle-stimulating hormone promotes RANK expression on human monocytes. *Cytokine.* 2011;53:141–144.
37. Cannon JG, Cortez-Cooper M, Meaders E, Stallings J, Haddow S, Kraj B, Sloan G, Mulloy A. Follicle-stimulating hormone, interleukin-1, and bone density in adult women. *Am J Physiol Regul Integr Comp Physiol.* 2010;298:R790–R798.
38. Gertz ER, Silverman NE, Wise KS, Hanson KB, Alekel DL, Stewart JW, Perry CD, Bhupathiraju SN, Kohut ML, Van Loan MD. Contribution of serum inflammatory markers to changes in bone mineral content and density in postmenopausal women: a 1-year investigation. *J Clin Densitom.* 2010;13:277–282.
39. Zhu LL, Blair H, Cao J, Yuen T, Latif R, Guo L, Tourkova IL, Li J, Davies TF, Sun L, Bian Z, Rosen C, Zallone A, New MI, Zaidi M. Blocking antibody to the beta-subunit of FSH prevents bone loss by inhibiting bone resorption and stimulating bone synthesis. *Proc Natl Acad Sci USA.* 2012;109:14574–14579.
40. Zhu LL, Tourkova I, Yuen T, Robinson LJ, Bian Z, Zaidi M, Blair HC. Blocking FSH action attenuates osteoclastogenesis. *Biochem Biophys Res Commun.* 2012;422:54–58.
41. Nishimori K, Young LJ, Guo Q, Wang Z, Insel TR, Matzuk MM. Oxytocin is required for nursing but is not essential for parturition or reproductive behavior. *Proc Natl Acad Sci USA.* 1996;93:11699–11704.
42. Young WS, 3rd. Shepard E, DeVries AC, Zimmer A, LaMarca ME, Ginns EI, Amico J, Nelson RJ, Hennighausen L, Wagner KU. Targeted reduction of oxytocin expression provides insights into its physiological roles. *Adv Exp Med Biol.* 1998;449:231–240.
43. Insel TR, Harbaugh CR. Lesions of the hypothalamic paraventricular nucleus disrupt the initiation of maternal behavior. *Physiol Behav.* 1989;45:1033–1041.
44. Mantella RC, Vollmer RR, Li X, Amico JA. Female oxytocin-deficient mice display enhanced anxiety-related behavior. *Endocrinology.* 2003;144:2291–2296.
45. Argiolas A, Collu M, Gessa GL, Melis MR, Serra G. The oxytocin antagonist d(CH2) 5Tyr(Me)-Orn8-vasotocin inhibits male copulatory behaviour in rats. *Eur J Pharmacol.* 1988;149:389–392.
46. Sclafani A, Rinaman L, Vollmer RR, Amico JA. Oxytocin knockout mice demonstrate enhanced intake of sweet and nonsweet carbohydrate solutions. *Am J Physiol Regul Integr Comp Physiol.* 2007;292:R1828–R1833.
47. Ferguson JN, Young LJ, Hearn EF, Matzuk MM, Insel TR, Winslow JT. Social amnesia in mice lacking the oxytocin gene. *Nat Genet.* 2000;25:284–288.
48. Copland JA, Ives KL, Simmons DJ, Soloff MS. Functional oxytocin receptors discovered in human osteoblasts. *Endocrinology.* 1999;140:4371–4374.
49. Colucci S, Colaianni G, Mori G, Grano M, Zallone A. Human osteoclasts express oxytocin receptor. *Biochem Biophys Res Commun.* 2002;297:442–445.

50. Colaianni G, Di Benedetto A, Zhu LL, Tamma R, Li J, Greco G, Peng Y, Dell'Endice S, Zhu G, Cuscito C, Grano M, Colucci S, Iqbal J, Yuen T, Sun L, Zaidi M, Zallone A. Regulated production of the pituitary hormone oxytocin from murine and human osteoblasts. *Biochem Biophys Res Commun.* 2011;411:512–515.
51. Tamma R, Colaianni G, Zhu LL, DiBenedetto A, Greco G, Montemurro G, Patano N, Strippoli M, Vergari R, Mancini L, Colucci S, Grano M, Faccio R, Liu X, Li J, Usmani S, Bachar M, Bab I, Nishimori K, Young LJ, Buettner C, Iqbal J, Sun L, Zaidi M, Zallone A. Oxytocin is an anabolic bone hormone. *Proc Natl Acad Sci USA.* 2009;106:7149–7154.
52. Elabd SK, Sabry I, Hassan WB, Nour H, Zaky K. Possible neuroendocrine role for oxytocin in bone remodeling. *Endocr Regul.* 2007;41:131–141.
53. Wysolmerski JJ. The evolutionary origins of maternal calcium and bone metabolism during lactation. *J Mammary Gland Biol Neoplasia.* 2002;7:267–276.
54. Sowers M, Eyre D, Hollis BW, Randolph JF, Shapiro B, Jannausch ML, Crutchfield M. Biochemical markers of bone turnover in lactating and nonlactating postpartum women. *J Clin Endocrinol Metab.* 1995;80:2210–2216.
55. Sun L, Tamma R, Yuen T, Colaianni G, Ji Y, Cuscito C, Bailey J, Dhawan S, Lu P, Calvano CD, Zhu LL, Zambonin CG, Di Benedetto A, Stachnik A, Liu P, Grano M, Colucci S, Davies TF, New MI, Zallone A, Zaidi M. Functions of vasopressin and oxytocin in bone mass regulation. *Proc Natl Acad Sci USA.* 2016;113:164–169.
56. Lotinun S, Limlomwongse L, Krishnamra N. The study of a physiological significance of prolactin in the regulation of calcium metabolism during pregnancy and lactation in rats. *Can J Physiol Pharmacol.* 1998;76:218–228.
57. Naylor KE, Iqbal P, Fledelius C, Fraser RB, Eastell R. The effect of pregnancy on bone density and bone turnover. *J Bone Miner Res.* 2000;15:129–137.
58. Lotinun S, Limlomwongse L, Sirikulchayanonta V, Krishnamra N. Bone calcium turnover, formation, and resorption in bromocriptine- and prolactin-treated lactating rats. *Endocrine.* 2003;20:163–170.
59. Meaney AM, Smith S, Howes OD, O'Brien M, Murray RM, O'Keane V. Effects of long-term prolactin-raising antipsychotic medication on bone mineral density in patients with schizophrenia. *Br J Psychiatry.* 2004;184:503–508.
60. Coss D, Yang L, Kuo CB, Xu X, Luben RA, Walker AM. Effects of prolactin on osteoblast alkaline phosphatase and bone formation in the developing rat. *Am J Physiol Endocrinol Metab.* 2000;279:E1216–E1225.
61. Seriwatanachai D, Thongchote K, Charoenphandhu N, Pandaranandaka J, Tudpor K, Teerapornpuntakit J, Suthiphongchai T, Krishnamra N. Prolactin directly enhances bone turnover by raising osteoblast-expressed receptor activator of nuclear factor kappaB ligand/osteoprotegerin ratio. *Bone.* 2008;42:535–546.
62. Seriwatanachai D, Charoenphandhu N, Suthiphongchai T, Krishnamra N. Prolactin decreases the expression ratio of receptor activator of nuclear factor kappaB ligand/osteoprotegerin in human fetal osteoblast cells. *Cell Biol Int.* 2008;32:1126–1135.
63. Krishnamra N, Seemoung J. Effects of acute and long-term administration of prolactin on bone 45Ca uptake, calcium deposit, and calcium resorption in weaned, young, and mature rats. *Can J Physiol Pharmacol.* 1996;74:1157–1165.

CHAPTER SIX

Gonadotrope Tumors

S.L. Asa*,†,1, S. Ezzat‡
* Department of Laboratory Medicine and Pathobiology, University of Toronto, Toronto, ON, Canada
† Department of Pathology, Laboratory Medicine Program, University Health Network, Toronto, ON, Canada
‡ Department of Medicine, University of Toronto, Endocrine Oncology, Princess Margaret Cancer Centre, University Health Network, Toronto, ON, Canada
[1] Corresponding author. E-mail address: sylvia.asa@uhn.ca

Contents

Abstract

Gonadotrope tumors arise from the gonadotropes of the adenohypophysis. These cells rarely give rise to hyperplasia, usually only in the setting of long-standing premature gonadal failure. In contrast, gonadotrope tumors represent one of the most frequent types of pituitary tumors. Despite their relatively common occurrence, the pathogenesis of gonadotrope tumors remains unknown. Effective nonsurgical therapies remain out of reach. We review the pituitary gonadotrope from the morphologic and functional perspectives to better understand its involvement as the cell of origin of a frequent type of pituitary tumor.

Progress in Molecular Biology and Translational Science, Volume 143
ISSN 1877-1173
http://dx.doi.org/10.1016/bs.pmbts.2016.08.005

1. INTRODUCTION

Gonadotrope tumors represent approximately one-fourth of all pituitary tumors. Although they are composed of differentiated adenohypophysial cells that are capable of producing components of gonadotropin hormones, these tumors are usually hormonally inactive and they do not often secrete intact hormones. Only rarely do they present with clinical manifestations of gonadotropin hormone excess; instead they have an insidious onset, coming to medical attention usually when they are compressive lesions associated with headaches, visual field disturbances, and varying degrees of hypopituitarism.

2. GONADOTROPES IN THE NORMAL PITUITARY

Gonadotropes represent one of the several types of hormone-secreting adenohypophysial cells. They produce the gonadotropins, follicle-stimulating hormone (FSH), and luteinizing hormone (LH). They are scattered throughout the pars distalis, or anterior pituitary, where they constitute approximately 10% of adenohypophysial cells in adult human pituitaries.[1] In addition, they represent the main type of hormone-producing cell of the pars tuberalis that is wrapped around the infundibulum, or the pituitary stalk, in humans.[2]

Gonadotropes have variable granular cytoplasmic positivity with the periodic acid-schiff (PAS) stain as well as with the basic aldehyde thionin and aldehyde fuchsin stains, however, these are not specific for gonadotropins and are therefore not reliable for cellular identification. Immunohistochemistry is the most reliable method to identify these cells. Gonadotropes exhibit nuclear reactivity for steroidogenic factor-1 (SF-1) and variable positivity for estrogen receptor-alpha (ERα). They contain cytoplasmic immunostaining for the common α-subunit of glycoprotein hormones, as well as the unique β-subunits of FSH and LH (Fig. 1). Both gonadotropins can be localized in a single cell, but there are differences in the numbers of cells containing each gonadotropin. Using double staining and immunoelectron microscopy, it has been shown that some gonadotropes express only one hormone at any given time, while others express both. The ultrastructural appearance of gonadotropes is characterized by polarity that is rather unique to these adenohypophysial cells; while they

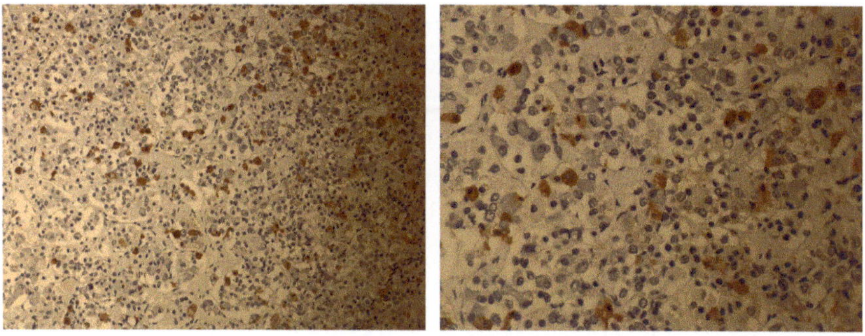

(A) (B)

Figure 1 Normal human gonadotropes. (A) At low magnification, gonadotropes containing β-LH are identified scattered throughout the acini of adenohypophysis among the other cell types that comprise this tissue. (B) Higher magnification of the same area identifies gonadotropes as round to polygonal cells with variable degrees of immunoreactivity for β-LH scattered throughout the cytoplasm.

are polyhedral, they are usually elongated in one direction and the nucleus usually is found at the pole of the cell that is closest to a basement membrane.[3,4] The cytoplasm contains numerous short profiles of slightly dilated rough endoplasmic reticulum; these dilated sacculi contain flocculent electron-lucent material. The cells have variably developed globoid Golgi complexes and mature secretory granules that are of heterogeneous morphology. The secretory granules of gonadotropes in men are small, averaging 250 nm, whereas those in gonadotropes of women can be as large as 300–600 nm. Both β-FSH and β-LH can be localized within the same secretory granules. The importance of paracrine interactions in the regulation of gonadotropes is highlighted by their intimate relationship with lactotrophs; the latter surround gonadotropes with elongated cell processes that form intercellular junctions between the two cell types.[3]

Gonadotropes are the only adenohypophysial cell type to exhibit sexual dimorphism during human fetal development.[5] Between 15 and 25 weeks of gestation, the pituitaries of female fetuses contain more gonadotropes than do those of male fetuses, and females have approximately equal numbers of cells with reactivity for β-FSH and β-LH, whereas in males, β-LH predominates. This dimorphism has been attributed to differences in hypothalamic GnRH levels.[6] In the adult human pituitary, there is no evidence of such dimorphism, unlike in rats where gonadotrope numbers and hormone content vary with age, sex, and hormonal status.[3] The gonadotropes of the human pars tuberalis undergo squamous metaplasia with aging.[2]

Gonadotropes are subject to alterations based on pathological changes in the hormonal milieu. Pharmacologic doses of estrogen cause a reduction in size, number, and hormone immunoreactivity of these cells[7]; similarly, during pregnancy they show involution.[8] Kallmann's syndrome is characterized by a reduction in the gonadotrope population.[9] In contrast, castration results in hypertrophy and hyperplasia of gonadotropes.

3. GONADOTROPE HYPERPLASIA

Gonadotrope hyperplasia is known to occur in patients with long-standing primary hypogonadism.[10,11] The hyperplastic cells develop an unusual morphology, leading to their descriptive names "gonadectomy cells," "gonadal deficiency cells," or "castration cells." The morphology can be recognized on routine histology; they are very large cells with abundant chromophobic vacuolated cytoplasm that contains PAS-positive granules. The vacuolation of the cytoplasm is due to accumulation of dilated endoplasmic reticulum that contains flocculent material; this cytoplasm can sometimes push the nucleus to the periphery, creating a "signet ring" cell appearance.[3] These cells stain weakly but diffusely for the α- and β-subunits of both gonadotropins, FSH and LH.

4. GONADOTROPE TUMORS

Pituitary tumors composed of adenohypophysial cells are classified as benign unless they give rise to distant metastases, which occurs only exceptionally rarely. Despite this benign categorization, these tumors exhibit a wide range of behaviors. Some are small and slowly growing, with virtually no hormonal hypersecretion; this type of adenohypophysial endocrine tumor is usually detected incidentally on radiologic examination performed for another reason. When they produce hormones in excess, small adenohypophysial tumors may be diagnosed because of the clinical endocrine syndrome they cause, but this is unusual among gonadotrope tumors. In contrast, hormonally inactive tumors usually present when they are large and invasive, giving rise to symptoms of an intracranial mass. The vast majority of clinically nonfunctioning pituitary adenomas and almost one-third of surgically resected tumors are now recognized to be gonadotrope tumors.[12–15]

4.1 Epidemiology

Adenohypophysial tumors were once considered to be rare, but with increasing awareness and more thorough data collection, it has become evident that they occur in almost 20% of the population.[16] They occur equally in both sexes and likely increase with age. Autopsy studies have reported pituitary tumors in up to 27% of patients. Clinical studies have documented prevalences of pituitary adenomas of 94 ± 19.3 cases per 100,000 people in Belgium,[17] 77.6/100,000 in the United Kingdom,[18] 3.9/100,000 in Sweden,[19] 115.57/100,000 in Iceland,[20] and 80.5/100,000 in Switzerland.[21] Nevertheless, the data from many countries are not available because they are not reported as cancer statistics.

The relative frequency of the various adenoma types and specifically of gonadotrope tumors is unknown because of incomplete data collection. Adenohypophysial endocrine tumors causing hyperprolactinemia are thought to be the most common type overall; the majority of these are treated medically and their incidence is unknown. However, among surgically resected tumors, one-third to one-quarter are classified as clinically nonfunctioning since they are not associated with hormone excess[22–27] and the majority of these are gonadotrope tumors.[28–32] Although there may be a female preponderance of pituitary tumors in some series, women usually present at a younger age with prolactin- and ACTH-secreting tumors whereas men tend to present later with clinically nonfunctioning gonadotrope tumors.[23] Gonadotrope adenomas are exceptionally uncommon in children.[33–35]

4.2 Clinical Presentation

Small gonadotrope tumors are usually identified incidentally at the time of investigation of other disorders. Large pituitary tumors cause symptoms of an intracranial mass, including headaches, nausea, and vomiting, and ultimately they give rise to visual field disturbances when they compress the optic chiasm; they can also invade laterally into the cavernous sinuses, upward into the parenchyma of the brain, and downward into the paranasal sinuses. Some patients present with pituitary apoplexy due to acute hemorrhage within the macroadenoma. Nevertheless, unlike inflammatory lesions and metastatic malignancies, they rarely result in diabetes insipidus or other cranial nerve palsies.

4.3 Biochemical Findings

Gonadotrope tumors are usually clinically nonfunctioning since they very rarely are associated with elevated gonadotropin levels.[12,14,36] The small group of clinically functioning gonadotrope tumors that cause elevations of circulating intact FSH and/or LH occur mainly in middle-aged men with a history of normal gonadal function.[14] It has been suggested that this sexual dimorphism is the due to the physiologic elevation of gonadotropins in peri- and postmenopausal women that masks both biochemical and functional evidence of tumoral hypersecretion. In contrast, men do not have high circulating gonadotropins unless they have primary gonadal failure, therefore, elevated gonadotropin levels raise the possibility of a pituitary gonadotrope tumor. Gonadotrope tumors only rarely occur in younger patients when they may be associated with primary hypogonadism.[11]

The patients may have hypopituitarism, often with hyperprolactinemia attributed to pituitary stalk compression. Hypopituitarism may involve only GH, or may also feature gonadotropin insufficiency.[12,37] Reduced gonadotropin levels may be the result of hyperprolactinemia rather than tissue destruction. Hypothyroidism or adrenocortical insufficiency are manifest less often as basal levels can be normal, but dynamic testing may elicit impaired corticotroph or thyrotroph responses.

It is paradoxical that gonadotrope tumors are usually associated with gonadal hypofunction. It is only very rarely that there is clinical and biochemical evidence of gonadal stimulation.[38] In young women, persistent elevation of serum gonadotropins causes inhibition of ovarian function that is reversible after tumor resection.[39]

In patients with functioning gonadotrope tumors, FSH levels are usually elevated, however, some patients have elevated levels of both FSH and LH. Only very rarely do these tumors produce LH excess alone.[40] Elevation of one gonadotropin alone with normal levels of the other or elevated gonadotropins in a patient with hypopituitarism suggests the diagnosis of a gonadotrope tumor, even in postmenopausal women. In some patients, circulating free α- and β-subunits have been documented, and the anomalous secretion of uncombined subunits by these tumors was confirmed in vitro.[41]

Dynamic testing can be used to establish the diagnosis of a gonadotropin-producing adenoma. Gonadotropins respond to stimulation by GnRH and paradoxically also respond to TRH stimulation. Both FSH and LH can show a paradoxical response, but it has been suggested that the FSH response involves the intact molecule whereas LH usually responds to TRH stimulation with release of the isolated β-subunit.[14,42–44] The value of this in routine practice is unclear. However, dynamic testing is more critical to evaluate other pituitary hormone reserves to determine the need for replacement therapy.

Gonadotropin excess in children is almost never due to a pituitary adenoma. Gonadotropin-dependent precocious puberty is most often due to germ cell tumors that produce chorionic gonadotropins rather than pituitary gonadotropins. Occasionally, hypothalamic hamartomas produce GnRH, causing precocious puberty. Other central nervous system tumors or tumor-like lesions can interfere with physiologic suppression of the hypothalamic-pituitary-gonadotropic axis in children. Excess circulating hCG secreted by choriocarcinoma or other germ cell tumors can be distinguished from pituitary gonadotropin hypersecretion using specific immunoassays.

4.4 Radiologic Features

Neuroradiologic examination classifies these tumors based on size and degree of local invasion. The original Hardy classification[45] defined *Grade I adenomas, or microadenomas*, as intrapituitary lesions that measure less than 1 cm in diameter; *Grade II adenomas* are intrasellar tumors larger than 1 cm in diameter (macroadenomas) or may have suprasellar extension without invasion; *Grade III adenomas* are locally invasive neoplasms; and *Grade IV adenomas* are large invasive neoplasms involving bone, hypothalamus, and/or the cavernous sinuses. This classification was modified by Wilson[46] to refine the classification of invasion further. The Hardy–Wilson classification was even further refined by Knosp[47] who classified parasellar growth into five grades. Gross total resection correlates inversely with grade, however, Grade 3 tumors were subsequently further divided into Grades 3A and 3B to distinguish the different resectability of tumors invading the superior cavernous sinus from those invading the inferior cavernous sinus.[48] Infiltrative pituitary tumors which invade dura, bone, and the cavernous sinuses are relatively common.[49–51] Large invasive pituitary tumors can invade the sphenoid bone downward to present as nasopharyngeal masses[52] or may invade posteriorly to destroy the clivus.[53]

Gonadotrope adenomas are usually Grade 3 or 4 lesions (Fig. 2A) that are large and extend upward into the suprasellar space, sometimes laterally into the cavernous sinuses and around the sella. Rarely, they are detected when they are small intrasellar lesions (Fig. 2B).

The radiologic classification is critical to determine surgical management. However, dural invasion increases with tumor size,[54,55] and invasion may be identified at surgery when it was not detected on preoperative

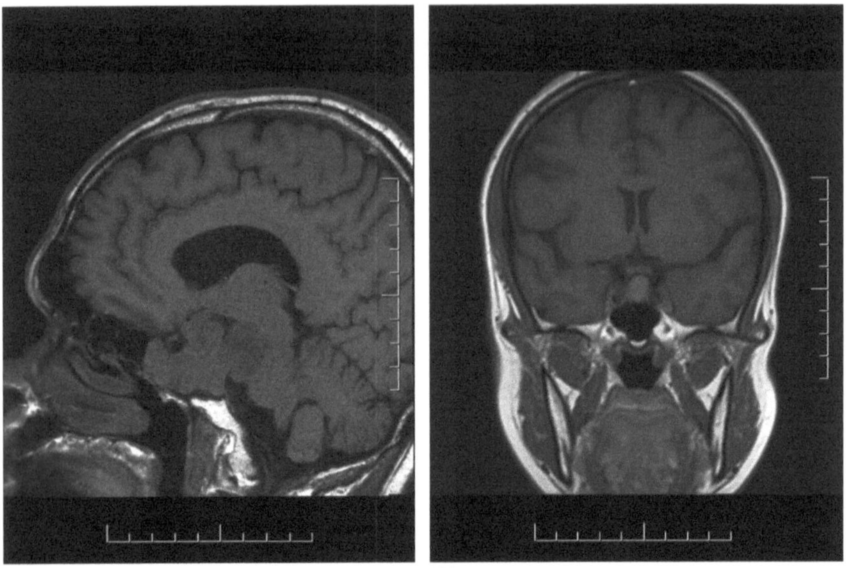

(A) (B)

Figure 2 Radiologic features of gonadotrope tumors. (A) These tumors usually present when they are large and invasive as in this macroadenoma that extends outside the sella turcica in all directions. (B) Rarely, a gonadotrope adenoma is identified when still small as in this example; this finding is usually identified during radiologic investigation for other reasons.

imaging.[54,55] Progress in imaging technology has improved preoperative assessment of pituitary tumors.[56,57] Sophisticated imaging reconstruction technologies allow determination of preoperative growth rates that predict postoperative growth of residual disease.[58]

Pituitary tumors including gonadotrope adenomas can develop ectopically in the sphenoid sinus and in parapharyngeal sites that are consistent with origin in adenohypophysial remnants of Rathke's cleft.[59–64] They have also been reported in suprasellar locations,[65,66] clivus,[67] hypothalamus,[68,69] and even in the third ventricle,[70] suggesting the possibility of excessive migration of Rathke's pouch in some individuals.

4.5 Morphology

Gonadotrope tumors are usually macroadenomas, large soft tumors that may have focal hemorrhage or necrosis because they outgrow their blood supply and grow in a hypoxic environment with reduced vascular density[71] that is recognized radiologically by the lack of contrast enhancement that identifies the nontumorous gland.

Gonadotrope tumors have a unique histologic appearance characterized by solid nests, trabecula, papillae, and pseudorosettes around blood vessels (Fig. 3A–B). The cells vary from chromophobic to pale acidophilic using conventional stains; this variability is due to partial or total oncocytic change that results in a granular acidophilic or amphophilic cytoplasm (Fig. 3C). When focal, oncocytic change increases with tumor recurrence.[72] Other adenohypophysial endocrine tumor types can be chromophobic or oncocytic, so this is not diagnostic. The PAS stain decorates scattered cytoplasmic granules, however, histochemical stains are nonspecific and cannot reliably

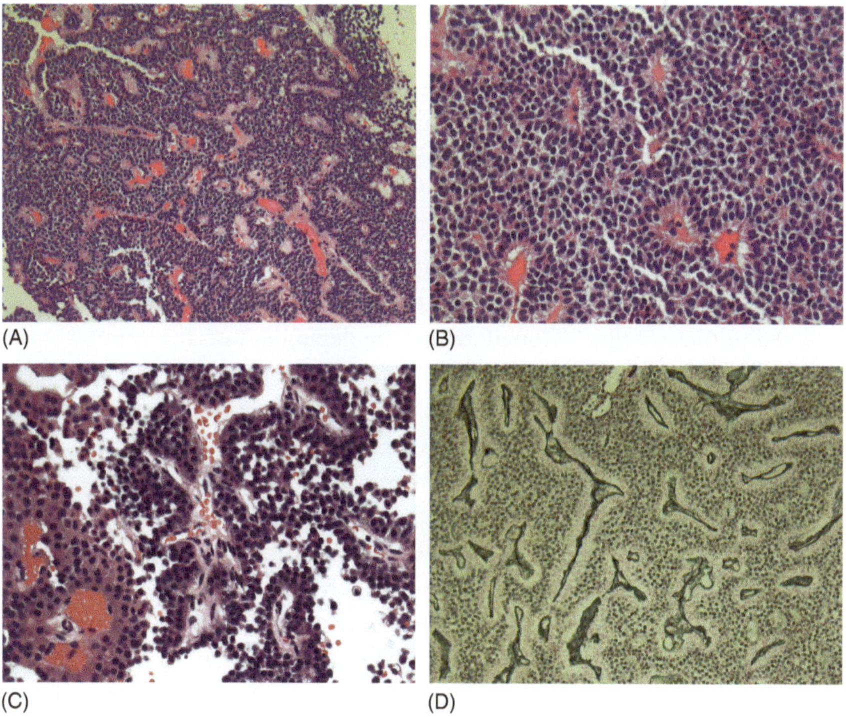

Figure 3 Histology of gonadotrope tumors. (A) Gonadotrope tumors are characterized by solid growth interspersed with trabecula and papillae with formation of prominent pseudorosettes around blood vessels. (B) The cells are mainly round or polygonal with chromophobic cytoplasm, but at this higher magnification, the polarity of the cells forming pseudorosettes around vascular channels can be appreciated. (C) The cell cytoplasm may show partial or total oncocytic change which results in a granular acidophilic or amphophilic cytoplasm. (D) The reticulin stain documents total breakdown of the normal reticulin that defines the acinar architecture of the normal adenohypophysis.

confirm the diagnosis. The diagnosis of adenoma is confirmed using a reticulin stain that documents total breakdown of the reticulin fiber network (Fig. 3D). The histologic diagnosis of a gonadotrope tumor is not possible without the use of immunohistochemistry.

Immunohistochemistry requires proper tissue fixation and the application of highly specific antibodies. The specificity of the reagents is of particular importance because of significant cross-reactivity attributable to the common α-subunit of the three glycoprotein hormones, thyrotropin (TSH) and the two gonadotropins, FSH and LH.[73] The α-subunit is also highly expressed by somatotrophs.[74,75] Therefore, it is recommended that laboratories use monoclonal antibodies that recognize the β-subunits of these glycoprotein hormones, providing higher specificity. The issue of tissue fixation is important because glycoproteins are particularly susceptible to degradation and conformational change due to poor fixation. Such fixation then alters antigen recognition sites and causes artefactual negativity.

The discovery of transcription factors that direct cytodifferentiation of adenohypophysial cells dramatically changed the diagnosis of these tumors. These markers allowed the recognition of cell differentiation even in the absence of hormone immunoreactivity. This has reduced the proportion of "null" cell tumors that were initially thought to be composed of undifferentiated pituitary cells.[72] Gonadotropes are dependent on the expression of SF-1 and GATA-2 as well as ER α.[1,31,32,76] In diagnostic labs, SF-1 is a valuable marker of these tumors that show diffuse strong nuclear staining (Fig. 4A).[1,32] In contrast, ERα positivity tends to be focal and weak (Fig. 4B).[1] In actively synthesizing tumors, the cytoplasm contains variable positivity for α-subunit, β-FSH, and β-LH; FSH (Fig. 4C) is usually more abundant than LH (Fig. 4D). However, many tumors exhibit only SF-1 immunoreactivity as a reliable marker of differentiation.

Electron microscopy (EM) played an important role in early studies of the classification of pituitary cells.[23] But, EM is not required for the majority of pituitary tumors in the era of sophisticated immunohistochemistry.[1] If performed, EM identifies tumor cells with variable degrees of differentiation. Well-differentiated gonadotropes are polar cells with euchromatic nuclei at the basal pole and abundant, well-differentiated cytoplasm that contains short profiles of dilated rough endoplasmic reticulum with flocculent, electron-lucent contents. Golgi complexes may be poorly developed or appear as large globular structures. Secretory granules are variable in number; they are most often small, measuring less than 250 nm in diameter, and moderately electron dense. In well-differentiated cells, the granules are

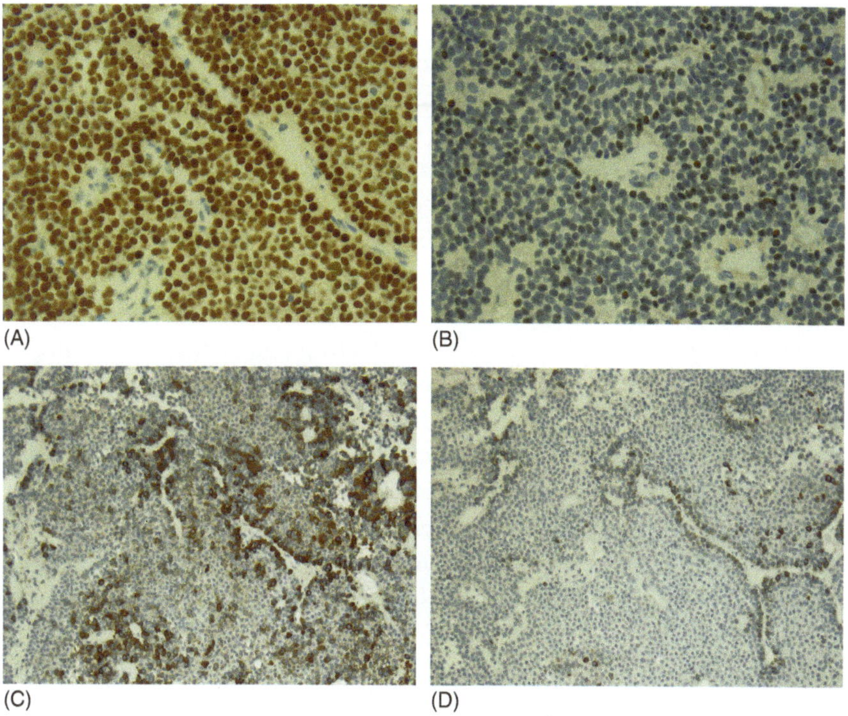

Figure 4 Immunoprofile of gonadotrope tumors. (A) The tumor cells exhibit strong and diffuse reactivity for steroidogenic factor-1 (SF1). (B) There is focal and variable positivity for estrogen receptor-α (ERα). (C) Gonadotrope tumors usually exhibit diffuse positivity for α-subunit of glycoprotein hormones (not shown) and focal staining for β-FSH (shown). (D) While there is strong staining for β-FSH in most gonadotrope tumors, the positivity for β-LH (shown) is usually less intense and found in fewer cells.

preferentially localized at the pole opposite the nucleus. Poorly differentiated cells tend to be polygonal or round with central nuclei and scant cytoplasmic organelles. Oncocytic change, characterized by increased numbers of dilated mitochondria, can be seen focally or throughout the majority of tumor cells.

In the rare patients with underlying prolonged primary hypogonadism, the nontumorous pituitary, if obtained for examination, contains prominent hyperplastic "gonadectomy" or "gonadal-deficiency" cells as described earlier.

An unusual tumor that was initially reported as "female-type gonadotroph adenoma"[77] is now no longer classified as such. Instead, the presence of a highly distinctive "honeycomb Golgi complex" is considered to be a feature of a variant of silent corticotroph adenoma.[78–80]

4.6 Prognostic and Predictive Markers

The 2004 WHO classification separated pituitary adenomas into "typical" and "atypical" variants[81] based on high mitotic index, a Ki-67 labeling index >3%, and nuclear reactivity for p53. However, these features were determined based on only a few studies and are neither widely accepted nor reproducible.[82] High mitotic indices and extensive nuclear p53 reactivity are not characteristic of pituitary tumors. The methodology used for determining a Ki-67 labeling index (Fig. 5) is variable.[83]

Indeed, the most important determinant of recurrence is surgical resectability as determined by radiology.[83] Gonadotroph adenomas express dopamine receptors,[84,85] and bromocriptine suppresses gonadotropin release in vitro[42,44,86] and in vivo.[86–88] Although some investigators promote the use of dopamine agonists for the management of residual disease,[89,90] it has no effect on tumor size, unlike the shrinkage seen in prolactinomas. Somatostatin analogs may reduce gonadotropin levels for the few patients with hormone excess[91,92] but they do not reduce tumor size effectively.[90] None of these features provide a biomarker that is helpful for prognosis or prediction.

4.7 Management Approaches

Patients with hormonally inactive pituitary tumors usually present relatively late when they have large mass lesions that require surgical decompression.[12] Due to the large differential diagnosis of sellar masses, the pathology diagnosis is critical to determine further therapeutic approaches. Complete

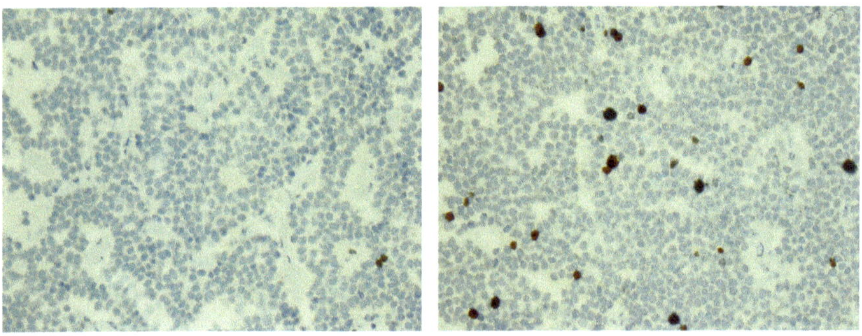

(A) (B)

Figure 5 Prognostic biomarkers of gonadotrope tumors. (A) The most widely used prognostic biomarker is Ki-67 that is labeled with the MIB-1 antibody. Most gonadotrope tumors have very low labeling indices as shown in this example. (B) Rarely gonadotrope tumors will exhibit focal increased labeling for Ki-67 as in this illustration.

surgical resection is the optimal goal, providing the most likely hope of cure,[45,93–108] however, even when incomplete, surgery alleviates symptoms and often results in improved pituitary function.

Recurrence within 10 years after surgery is not uncommon, reaching 50%.[12] Nevertheless, repeat resection is feasible and alleviates symptoms. The use of dopamine agonists or somatostatin analogs has been suggested to be helpful in restraining tumor growth.[89,90]

Radiation therapy has been recommended liberally in the past and continues to be used in some centers. However, many groups now reserve this treatment for patients with more rapid and aggressive tumor regrowth,[12] a feature that is more frequent in other clinically nonfunctioning tumors than in true gonadotrope adenomas.[109] The recent development of peptide receptor radiotherapy to deliver targeted radiation through somatostatin binding to its receptors may prove useful for the most aggressive lesions.[110]

4.8 Pathogenesis

The vast majority of gonadotrope tumors are sporadic neoplasms whose etiology is unknown. They rarely arise in the setting of *multiple endocrine neoplasia (MEN)* syndromes that include pituitary tumors. The most common of these is MEN-1, an autosomal dominant disorder with incomplete penetrance due to inheritance of a mutated MEN1 gene on chromosome 11q13 that encodes the tumor suppressor protein "menin".[111] Loss of menin expression by mutation and LOH or through epigenetic silencing is not found in sporadic adenomas.[112,113] Another variant of MEN is MEN-4, due to germline mutations in the cyclin-dependent kinase inhibitors (CDKIs) CDKNIB/p27Kip1[114,115] and CDKN2C/p18INK4c,[116] but again, the products of these genes, p27 and p18 are not usually downregulated in sporadic tumors. Gonadotrope tumors are not characteristic of *Carneys complex (CNC)*, another familial disorder with autosomal dominant inheritance due to germline mutations of the PRKAR1Aα gene encoding the PKA regulatory subunit 1Aα [117,118]; the PRKR1Aα is not mutated in sporadic pituitary tumors.[119,120]

Familial isolated pituitary adenoma (FIPA) syndrome may feature gonadotrope lesions; however, the majority of patients and families have lactotrope or somatotrope tumors.[121] About 15% of FIPA kindreds carry germline mutations in the aryl hydrocarbon receptor-interacting protein (AIP) gene and LOH is implicated in loss of the normal allele with reduced expression of AIP in inherited tumors.[121,122] AIP mutations are not frequent in sporadic pituitary tumors[123] and although epigenetic silencing of AIP has been reported in some somatotrope tumors,[124] this has not been identified in gonadotrope tumors.

Germline mutations in the succinate dehydrogenase genes (SDHA, SDHB, SDHC, and SDHD) are known to predispose to the development of pheochromocytoma/paraganglioma, renal carcinoma, and gastrointestinal stromal tumor. The occurrence of pituitary adenomas in some patients with this syndrome[125] raised the possibility of SDH mutation as a cause of sporadic pituitary adenomas, however, this has been found to be an exceptionally rare phenomenon.[126]

The etiologic factors underlying *sporadic pituitary tumors* include rare genetic events, hormonal stimulation, and growth factors.[1,127–129] A recent study reported whole exome sequencing of 7 pituitary gonadotrope tumors[130]; the authors identified 24 somatic variants in independent genes with no recurrent mutations. Mutations were not observed in genes that have been associated with pituitary tumorigenesis as discussed earlier. Somatic variants were found in platelet-derived growth factor D (PDGFD), N-myc downregulated gene family member 4 (NDRG4), and Zipper sterile-alpha-motif kinase (ZAK) but these are not implicated in pituitary tumor etiology. These findings suggest that mechanisms other than somatic mutation are involved in the etiology of these tumors.[130]

Growth factors regulate adenohypophysial expansion during development[131–134] and altered growth factor action has been implicated in pituitary tumorigenesis. Expression of transforming growth factor (TGF) α[135,136] is altered and there are reports suggesting a role for the epidermal growth factor (EGF) family and its receptors in pituitary tumorigenesis.[137–139] Inhibins, activins, and follistatin[140,141] have all been identified as putative dysregulated tumorigenic factors. The FGF family may also play a role in cell proliferation; expression of the FGF ligands is high[142] and the pattern of expression of their receptors (FGFRs) is altered in pituitary tumors compared with nontumorous adenohypophysis.[142] FGFR2 is expressed in normal adenohypophysial cells and it suppresses the cancer/testis-melanoma associated antigen (MAGE-A)[143] but this receptor is silenced by promoter methylation in pituitary tumors.[144] In contrast, FGFR4 is overexpressed in pituitary tumors.[142] A truncated isoform of FGFR4, ptd-FGFR4 is expressed in aggressive pituitary tumors[145]; its expression is thought to result from alternative transcription initiation at a cryptic intronic promoter[146] and this variant interrupts adenohypophysial cell adhesion by displacing N-cadherin from NCAM/N-cadherin/FGFR4 complexes at the cell membrane.[147] Similarly, PSA-NCAM has been implicated in pituitary tumor growth and invasiveness[148] and may also function by disrupting the tri-protein cell adhesion complexes.

Pituitary tumors have been described to have altered expression of the cell cycle proteins pRb, p21, and p27. Epigenetic regulators, such as the DNA methyltransferase (DNMT) enzymes, may be the etiologic factors silencing these tumor suppressors,[149] but there are also data suggesting increased degradation as a mechanism of p27 loss.[150] Epigenetic silencing of p27 may also be due to alterations in the expression of the microRNAs mi221 and miR222.[151] Mutation of DKC1, which encodes dyskerin, a pseudo-uridine synthase that modifies ribosomal RNA, has been reported in a single pituitary tumor associated with diminished p27 expression.[152] The transcription factor Ikaros is an important epigenetic regulator that is known to play a role in hematologic cell proliferation and neoplasia as well as pituitary development[153,154]; approximately half of pituitary tumors express a dominant negative isoform, Ik6[155] that causes enhanced cell survival through evasion of apoptosis by Bcl-XL activation.[156] Ikaros also promotes cholesterol uptake that is required for tumor cells metabolism, hormone secretion,[157] and survival in the hypoxic environment of pituitary tumors.[71,158] The high mobility group (HMG) proteins containing AT-hook domains (HMGA), another group of chromatin remodeling proteins, have also been implicated in pituitary tumorigenesis[159,160]; in mice, they have been shown to upregulate pituitary cyclin B2.[161] Human pituitary tumors have been reported to have amplification and overexpression of HMGA1 or 2,[161–163] and the dysregulation of HMGA2 has been associated with altered expression of the Let-7 microRNA that is thought to target this gene.[164]

These data indicate that the etiology of the most common sporadic pituitary adenomas is likely attributable to epigenetic signals that target growth factors and cell cycle regulators, rather than intragenic mutations that characterize most solid malignancies. Unfortunately, there are very few animal models of gonadotrope tumors[165] and the only cell line available[166] has proven to have restricted capacity for experimentation, limiting studies of these lesions.

5. CONCLUSIONS

The hypothalamic-pituitary-gonadal axis represents one of the most complex and intricately networked systems in mammals. The perpetual demands placed on this system during health and disease requires highly coordinated functions. It appears likely that stressful demands engage epigenetic mechanisms that, within the adenohypophysis, frequently involve gonadotropes. This positions gonadotrope tumors at the center stage as the most common form of pituitary neoplasia. Greater insights into the machinery of the supply and demand neuroendocrine network of reproduction will prove essential for the next wave of progress in this field.

REFERENCES

1. Asa SL. Tumors of the pituitary gland. In: Silverberg SG, ed. *AFIP Atlas of Tumor Pathology, Series 4, Fascicle 15. Silver Spring, MD: ARP Press; 2011.*
2. Asa SL, Kovacs K, Bilbao JM. The pars tuberalis of the human pituitary. A histologic, immunohistochemical, ultrastructural and immunoelectron microscopic analysis. *Virchows Arch A Pathol Anat Histopathol.* 1983;399:49–59.
3. Horvath E, Kovacs K. Fine structural cytology of the adenohypophysis in rat and man. *J Electron Microsc Tech.* 1988;8:401–432.
4. Horvath E, Kovacs K. The adenohypophysis. In: Kovacs K, Asa SL, eds. *Functional Endocrine Pathology. Boston: Blackwell Scientific Publications, Inc.; 1991:245–281.*
5. Asa SL, Kovacs K, Laszlo FA, Domokos I, Ezrin C. Human fetal adenohypophysis. Histologic and immunocytochemical analysis. *Neuroendocrinology.* 1986;43:308–316.
6. Asa SL, Kovacs K. Functional morphology of the human fetal pituitary. *Pathol Annu.* 1984;19(Pt 1):275–315.
7. Scheithauer BW, Kovacs K, Randall RV, Ryan N. Effects of estrogen on the human pituitary: a clinicopathologic study. *Mayo Clin Proc.* 1989;64:1077–1084.
8. Scheithauer BW, Sano T, Kovacs KT, Young Jr WF, Ryan N, Randall RV. The pituitary gland in pregnancy: a clinicopathologic and immunohistochemical study of 69 cases. *Mayo Clin Proc.* 1990;65:461–474.
9. Kovacs K, Sheehan HL. Pituitary changes in Kallman's syndrome: a histologic, immunocytologic, ultrastructural, and immunoelectron microscopic study. *Fertil Steril.* 1982;37:83–89.
10. Nicolis G, Shimshi M, Allen C, Halmi NS, Kourides IA. Gonadotropin-producing pituitary adenoma in a man with long-standing primary hypogonadism. *J Clin Endocrinol Metab.* 1988;66:237–241.
11. Kovacs K, Horvath E, Rewcastle NB, Ezrin C. Gonadotroph cell adenoma of the pituitary in a woman with long-standing hypogonadism. *Arch Gynecol.* 1980;229:57–65.
12. Jaffe CA. Clinically non-functioning pituitary adenoma. *Pituitary.* 2006;9(4):317–321.
13. Trouillas J, Girod C, Sassolas G, Claustrat B. The human gonadotropic adenoma: pathologic diagnosis and hormonal correlations in 26 tumors. *Semin Diagn Pathol.* 1986;3:42–57.
14. Snyder PJ. Gonadotroph cell adenomas of the pituitary. *Endocr Rev.* 1985;6:552–563.
15. Daneshdoost L, Gennarelli TA, Bashey HM, et al. Recognition of gonadotroph adenomas in women. *N Engl J Med.* 1991;324:589–594.
16. Ezzat S, Asa SL, Couldwell WT, et al. The prevalence of pituitary adenomas: a systematic review. *Cancer.* 2004;101(3):613–619.
17. Daly AF, Rixhon M, Adam C, Dempegioti A, Tichomirowa MA, Beckers A. High prevalence of pituitary adenomas: a cross-sectional study in the province of Liege, Belgium. *J Clin Endocrinol Metab.* 2006;91(12):4769–4775.
18. Fernandez A, Karavitaki N, Wass JA. Prevalence of pituitary adenomas: a community-based, cross-sectional study in Banbury (Oxfordshire, UK). *Clin Endocrinol (Oxf).* 2009;72:377–382.

19. Tjornstrand A, Gunnarsson K, Evert M, et al. The incidence rate of pituitary adenomas in western Sweden for the period 2001–2011. *Eur J Endocrinol.* 2014;171(4): 519–526.
20. Agustsson TT, Baldvinsdottir T, Jonasson JG, et al. The epidemiology of pituitary adenomas in Iceland, 1955–2012: a nationwide population-based study. *Eur J Endocrinol.* 2015;173(5):655–664.
21. Fontana E, Gaillard R. Epidemiology of pituitary adenoma: results of the first Swiss study. *Rev Med Suisse.* 2009;5(223):2172–2174.
22. Wilson CB, Dempsey LC. Transsphenoidal microsurgical removal of 250 pituitary adenomas. *J Neurosurg.* 1978;48:13–22.
23. Kovacs K, Horvath E. Tumors of the pituitary gland. *Atlas of Tumor Pathology, Second Series, Fascicle 21. Washington, DC: Armed Forces Institute of Pathology; 1986.*
24. Klibanski A, Zervas NT. Diagnosis and management of hormone-secreting pituitary adenomas. *N Engl J Med.* 1991;324:822–831.
25. Mindermann T, Wilson CB. Age-related and gender-related occurrence of pituitary adenomas. *Clin Endocrinol (Oxf).* 1994;41:359–364.
26. Terada T, Kovacs K, Stefaneanu L, Horvath E. Incidence, pathology, and recurrence of pituitary adenomas: study of 647 unselected surgical cases. *Endocr Pathol.* 1995;6: 301–310.
27. Feldkamp J, Santen R, Harms E, Aulich A, Modder U, Scherbaum WA. Incidentally discovered pituitary lesions: high frequency of macroadenomas and hormone-secreting adenomas—results of a prospective study. *Clin Endocrinol (Oxf).* 1999;51(1):109–113.
28. Asa SL, Gerrie BM, Singer W, Horvath E, Kovacs K, Smyth HS. Gonadotropin secretion in vitro by human pituitary null cell adenomas and oncocytomas. *J Clin Endocrinol Metab.* 1986;62:1011–1019.
29. Yamada S, Asa SL, Kovacs K. Oncocytomas and null cell adenomas of the human pituitary: morphometric and in vitro functional comparison. *Virchows Arch A Pathol Anat Histopathol.* 1988;413:333–339.
30. Yamada S, Asa SL, Kovacs K, Muller P, Smyth HS. Analysis of hormone secretion by clinically nonfunctioning human pituitary adenomas using the reverse hemolytic plaque assay. *J Clin Endocrinol Metab.* 1989;68:73–80.
31. Zafar M, Ezzat S, Ramyar L, Pan N, Smyth HS, Asa SL. Cell-specific expression of estrogen receptor in the human pituitary and its adenomas. *J Clin Endocrinol Metab.* 1995;80:3621–3627.
32. Asa SL, Bamberger A-M, Cao B, Wong M, Parker KL, Ezzat S. The transcription activator steroidogenic factor-1 is preferentially expressed in the human pituitary gonadotroph. *J Clin Endocrinol Metab.* 1996;81:2165–2170.
33. Kane LA, Leinung MC, Scheithauer BW, et al. Pituitary adenomas in childhood and adolescence. *J Clin Endocrinol Metab.* 1994;79:1135–1140.
34. Mukai K, Seljeskog EL, Dehner LP. Pituitary adenomas in patients under 20 years old. A clinicopathological study of 12 cases. *J Neurooncol.* 1986;4:79–89.
35. Jagannathan J, Dumont AS, Jane Jr JA, Laws Jr ER. Pediatric sellar tumors: diagnostic procedures and management. *Neurosurg Focus.* 2005;18(6A):E6.
36. Snyder PJ. Gonadotroph cell pituitary adenomas. *Endocrinol Metab Clin North Am.* 1987;16:755–764.
37. Klibanski A. Nonsecreting pituitary tumors. *Endocrinol Metab Clin North Am.* 1987;16: 793–804.
38. Djerassi A, Coutifaris C, West VA, Asa SL, Kapoor SC, Snyder PJ. Gonadotroph adenoma in a premenopausal woman secreting FSH and causing ovarian hyperstimulation. *J Clin Endocrinol Metab.* 1995;80:591–594.
39. Cook DM, Watkins S, Snyder PJ. Gonadotrophin-secreting pituitary adenomas masquerading as primary ovarian failure. *Clin Endocrinol (Oxf).* 1986;25:729–738.

40. Klibanski A, Deutsch PJ, Jameson JL, et al. Luteinizing hormone-secreting pituitary tumor: biosynthetic characterization and clinical studies. *J Clin Endocrinol Metab.* 1987;64:536–542.
41. Snyder PJ, Bashey HM, Kim SU, Chappel SC. Secretion of uncombined subunits of luteinizing hormone by gonadotroph cell adenomas. *J Clin Endocrinol Metab.* 1984;59:1169–1175.
42. Lamberts SWJ, Verleun T, Oosterom R, et al. The effects of bromocriptine, thyrotropin-releasing hormone, and gonadotropin-releasing hormone on hormone secretion by gonadotropin-secreting pituitary adenomas *in vivo* and *in vitro*. *J Clin Endocrinol Metab.* 1987;64:524–530.
43. Asa SL, Gerrie BM, Kovacs K, et al. Structure-function correlations of human pituitary gonadotroph adenomas *in vitro*. *Lab Invest.* 1988;58:403–410.
44. Kwekkeboom DJ, de Jong FH, Lamberts SWJ. Gonadotropin release by clinically nonfunctioning and gonadotroph pituitary adenomas *in vivo* and *in vitro*: relation to sex and effects of thyrotropin-releasing hormone, gonadotropin-releasing hormone, and bromocriptine. *J Clin Endocrinol Metab.* 1989;68:1128–1135.
45. Hardy J. Transsphenoidal surgery of hypersecreting pituitary tumors. In: Kohler PO, Ross GT, eds. *Diagnosis and Treatment of Pituitary Tumors. Int. Congress Series No. 303. Amsterdam: Exerpta Medica;1973:179–198.*
46. Wilson CB. Neurosurgical management of large and invasive pituitary tumors. In: Tindall GT, Collins WF, eds. *Clinical Management of Pituitary Disorders. New York: Raven Press;1979:335–342.*
47. Knosp E, Steiner E, Kitz K, Matula C. Pituitary adenomas with invasion of the cavernous sinus space: a magnetic resonance imaging classification compared with surgical findings. *Neurosurgery.* 1993;33(4):610–617.
48. Micko AS, Wohrer A, Wolfsberger S, Knosp E. Invasion of the cavernous sinus space in pituitary adenomas: endoscopic verification and its correlation with an MRI-based classification. *J Neurosurg.* 2015;122(4):803–811.
49. Scheithauer BW, Kovacs KT, Laws Jr ER, Randall RV. Pathology of invasive pituitary tumors with special reference to functional classification. *J Neurosurg.* 1986;65: 733–744.
50. Selman WR, Laws Jr ER, Scheithauer BW, Carpenter SM. The occurrence of dural invasion in pituitary adenomas. *J Neurosurg.* 1986;64:402–407.
51. Sautner D, Saeger W. Invasiveness of pituitary adenomas. *Pathol Res Pract.* 1991;187: 632–636.
52. van der Mey AG, van Seters AP, van Krieken JH, Vielvoye J, Van Dulken H, Hulshof JH. Large pituitary adenomas with extension into the nasopharynx. Report of three cases with a review of the literature. *Ann Otol Rhinol Laryngol.* 1989;98:618–624.
53. Wong K, Raisanen J, Taylor SL, McDermott MW, Wilson CB, Gutin PH. Pituitary adenoma as an unsuspected clival tumor. *Am J Surg Pathol.* 1995;19:900–903.
54. Meij BP, Lopes MB, Ellegala DB, Alden TD, Laws Jr ER. The long-term significance of microscopic dural invasion in 354 patients with pituitary adenomas treated with transsphenoidal surgery. *J Neurosurg.* 2002;96(2):195–208.
55. Lonser RR, Ksendzovsky A, Wind JJ, Vortmeyer AO, Oldfield EH. Prospective evaluation of the characteristics and incidence of adenoma-associated dural invasion in Cushing disease. *J Neurosurg.* 2012;116(2):272–279.
56. Hess CP, Dillon WP. Imaging the pituitary and parasellar region. *Neurosurg Clin N Am.* 2012;23(4):529–542.
57. Lucas JW, Zada G. Imaging of the pituitary and parasellar region. *Semin Neurol.* 2012;32 (4):320–331.
58. Monsalves E, Larjani S, Loyola GB, et al. Growth patterns of pituitary adenomas and histopathological correlates. *J Clin Endocrinol Metab.* 2014;99(4):1330–1338.

59. Rasmussen P, Lindholm J. Ectopic pituitary adenomas. *Clin Endocrinol (Oxf).* 1979;11:69–74.
60. Lloyd RV, Chandler WF, Kovacs K, Ryan N. Ectopic pituitary adenomas with normal anterior pituitary glands. *AmJ Surg Pathol.* 1986;108:546–552.
61. Shenker Y, Lloyd RV, Weatherbee L, Port FK, Grekin RJ, Barkan AL. Ectopic prolactinoma in a patient with hyperparathyroidism and abnormal sellar radiography. *J Clin Endocrinol Metab.* 1986;62:1065–1069.
62. Slonim SM, Haykal HA, Cushing GW, Freidberg SR, Lee AK. MRI appearances of an ectopic pituitary adenoma: case report and review of the literature. *Neuroradiol.* 1993;35:546–548.
63. Kikuchi K, Kowada M, Sasaki J, Sageshima M. Large pituitary adenoma of the sphenoid sinus and the nasopharynx: report of a case with ultrastructural evaluations. *Surg Neurol.* 1994;42:330–334.
64. Coire CI, Horvath E, Kovacs K, Smyth HS, Ezzat S. Cushing's syndrome from an ectopic pituitary adenoma with peliosis: a histological, immunohistochemical and ultrastructural study and review of the literature. *Endocr Pathol.* 1997;8:65–74.
65. Hori A. Suprasellar peri-infundibular ectopic adenohypophysis in fetal and adult brains. *J Neurosurg.* 1985;62:113–115.
66. Dyer EH, Civit T, Abecassis J-P, Derome PJ. Functioning ectopic supradiaphragmatic pituitary adenomas. *Neurosurgery.* 1994;43:529–532.
67. Anand NK, Osborne CM, Harkey HLI. Infiltrative clival pituitary adenoma of ectopic origin. *Head Neck Surg.* 1993;108:178–183.
68. Lindboe CF, Unsgard G, Myhr G, Scott H. ACTH and TSH producing ectopic suprasellar pituitary adenoma of the hypothalamic region: case report. *Clin Neuropathol.* 1993;12:138–141.
69. Matsumura A, Meguro K, Doi M, Tsurushima H, Tomono Y. Suprasellar ectopic pituitary adenoma: case report and review of the literature. *Neurosurgery.* 1990;26: 681–685.
70. Kleinschmidt-De Masters BK, Winston KR, Rubinstein D, Samuels MH. Ectopic pituitary adenomas of the third ventricle. Case report. *J Neurosurg.* 1990;72:139–142.
71. Vidal S, Kovacs K, Horvath E, Scheithauer BW, Kuroki T, Lloyd RV. Microvessel density in pituitary adenomas and carcinomas. *Virchows Arch A Pathol Anat Histopathol.* 2001;438(6):595–602.
72. Kovacs K, Horvath E, Ryan N, Ezrin C. Null cell adenoma of the human pituitary. *Virchows Arch A Pathol Anat Histopathol.* 1980;387:165–174.
73. Labat-Moleur F, Trouillas J, Seret-Begue D, Kujas M, Delisle M-B, Ronin C. Evaluation of 29 monoclonal and polyclonal antibodies used in the diagnosis of pituitary adenomas. A collaborative study from pathologists of the Club Français de l'Hypophyse. *Pathol Res Pract.* 1991;187:534–538.
74. Kontogeorgos G, Kovacs K, Scheithauer BW, Rologis D, Orphanidis G. a-Subunit immunoreactivity in plurihormonal pituitary adenomas of patients with acromegaly. *Mod Pathol.* 1991;4:191–195.
75. Kontogeorgos G, Asa SL, Kovacs K, Smyth HS, Singer W. Production of alpha-subunit of glycoprotein hormones by pituitary somatotroph adenomas in vitro. *Acta Endocrinol (Copenh).* 1993;129:565–572.
76. Friend KE, Chiou YK, Lopes MBS, Laws Jr ER, Hughes KM, Shupnik MA. Estrogen receptor expression in human pituitary: correlation with immunohistochemistry in normal tissue, and immunohistochemistry and morphology in macroadenomas. *J Clin Endocrinol Metab.* 1994;78:1497–1504.
77. Horvath E, Kovacs K. Gonadotroph adenomas of the human pituitary: sex-related fine-structural dichotomy. A histologic, immunocytochemical, and electron-microscopic study of 30 tumors. *Am J Pathol.* 1984;117:429–440.

78. Sano T, Kovacs K, Asa SL, et al. Pituitary adenoma with "honeycomb Golgi" appearance showing a phenotypic change at recurrence from clinically nonfunctioning to typical Cushing disease. *Endocr Pathol.* 2002;13(2):125–130.
79. Sano T, Mader R, Asa SL, Qian ZR, Hino A, Yamada S. "Honeycomb Golgi" in pituitary adenomas: not a marker of gonadotroph adenomas. *Endocr Pathol.* 2003;14 (4):363–368.
80. Kim K, Yamada S, Usui M, Sano T. Co-localization of honeycomb golgi and ACTH granules in a giant ACTH-producing pituitary adenoma. *Endocr Pathol.* 2005;16 (3):239–244.
81. DeLellis RA, Lloyd RV, Heitz PU, Eng C. *Pathology and Genetics of Tumours of Endocrine Organs.* Lyons, France: IARC Press; 2004.
82. Al Shraim M, Asa SL. The 2004 World Health Organization classification of pituitary tumors: what is new? *Acta Neuropathol.* 2006;111(1):1–7.
83. Asa SL, Ezzat S. Aggressive pituitary tumors or localized pituitary carcinomas: defining pituitary tumors. *Expert Rev Endocrinol Metab.* 2016;11(2):149–162.
84. Koga M, Nakao H, Arao M, et al. Demonstration of specific dopamine receptors on human pituitary adenomas. *Acta Endocrinol (Copenh).* 1987;114:595–602.
85. Lloyd RV, Anagnostou D, Chandler WF. Dopamine receptors in immunohistochemically characterized null cell adenomas and normal human pituitaries. *Mod Pathol.* 1988;1:51–56.
86. Kwekkeboom DJ, Hofland LJ, van Koetsveld PM, Singh R, van den Berge JH, Lamberts SWJ. Bromocriptine increasingly suppresses the *in vitro* gonadotropin and a-subunit release from pituitary adenomas during long term culture. *J Clin Endocrinol Metab.* 1990;71:718–724.
87. Vance ML, Ridgway EC, Thorner MO. Follicle-stimulating hormone- and a-subunit-secreting pituitary tumor treated with bromocriptine. *J Clin Endocrinol Metab.* 1985;61:580–584.
88. Klibanski A, Shupnik MA, Bikkal HA, Black PM, Kliman B, Zervas NT. Dopaminergic regulation of a-subunit secretion and messenger ribonucleic acid levels in a-secreting pituitary tumors. *J Clin Endocrinol Metab.* 1988;66:96–102.
89. Bevan JS, Webster J, Burke CW, Scanlon MF. Dopamine agonists and pituitary tumor shrinkage. *Endocr Rev.* 1992;13(2):220–240.
90. Chanson P, Brochier S. Non-functioning pituitary adenomas. *J Endocrinol Invest.* 2005;28(11 Suppl International):93–99.
91. Lamberts SWJ. The role of somatostatin in the regulation of anterior pituitary hormone secretion and the use of its analogs in the treatment of human pituitary tumors. *Endocr Rev.* 1988;9:417–436.
92. Vos P, Croughs RJM, Thijssen JHH, van't Verlaat JW, van Ginkel LA. Response of luteinizing hormone secreting pituitary adenoma to a long-acting somatostatin analogue. *Acta Endocrinol (Copenh).* 1988;118:587–590.
93. Kelly DF. Transsphenoidal surgery for Cushing's disease: a review of success rates, remission predictors, management of failed surgery, and Nelson's Syndrome. *Neurosurg Focus.* 2007;23(3):E5.
94. Oyesiku NM. Stereotactic radiosurgery for Cushing disease: a review. *Neurosurg Focus.* 2007;23(6):E14.
95. Nomikos P, Buchfelder M, Fahlbusch R. The outcome of surgery in 668 patients with acromegaly using current criteria of biochemical 'cure'. *Eur J Endocrinol.* 2005;152 (3):379–387.
96. Hammer GD, Tyrrell JB, Lamborn KR, et al. Transsphenoidal microsurgery for Cushing's disease: initial outcome and long-term results. *J Clin Endocrinol Metab.* 2004;89(12):6348–6357.

97. Beauregard C, Truong U, Hardy J, Serri O. Long-term outcome and mortality after transsphenoidal adenomectomy for acromegaly. *Clin Endocrinol (Oxf).* 2003;58 (1):86–91.
98. Kreutzer J, Vance ML, Lopes MB, Laws Jr ER. Surgical management of GH-secreting pituitary adenomas: an outcome study using modern remission criteria. *J Clin Endocrinol Metab.* 2001;86(9):4072–4077.
99. Sheaves R, Jenkins P, Blackburn P, et al. Outcome of transsphenoidal surgery for acromegaly using strict criteria for surgical cure. *Clin Endocrinol (Oxf).* 1996;45 (4):407–413.
100. Ross DA, Wilson CB. Results of transphenoidal microsurgery for growth hormone-secreting pituitary adenoma in a series of 214 patients. *J Neurosurg.* 1988;68:854–867.
101. Roelfsema F, Van Dulken H, Frölich M. Long-term results of transsphenoidal pituitary microsurgery in 60 acromegalic patients. *Clin Endocrinol (Oxf).* 1985;23:555–565.
102. Hardy J. Transsphenoidal microsurgery of prolactinomas. In: Black PM, Zervas N, Ridgeway EC, eds. et al. *Secretory Tumors of the Pituitary Gland. New York: Raven Press; 1984:73–81.*
103. Ciric I, Mikhael M, Stafford T, Lawson L, Garces R. Transsphenoidal microsurgery of pituitary macroadenomas with long-term follow-up results. *J Neurosurg.* 1983;59: 395–401.
104. Baskin DS, Boggan JE, Wilson CB. Transsphenoidal microsurgical removal of growth hormone-secreting pituitary adenomas. A review of 137 cases. *J Neurosurg.* 1982;56: 634–641.
105. Domingue JN, Richmond IL, Wilson CB. Results of surgery in 114 patients with prolactin-secreting pituitary adenomas. *Am J Obstet Gynecol.* 1980;137:102–108.
106. Tindall GT, McLanahan CS, Christy JH. Transsphenoidal microsurgery for pituitary tumors associated with hyperprolactinemia. *J Neurosurg.* 1978;48:849.
107. Giovanelli MA, Motti EDF, Paracchi A, Beck-Peccoz P, Ambrosi B, Faglia G. Treatment of acromegaly by transsphenoidal microsurgery. *J Neurosurg.* 1976;44: 677–686.
108. Hardy J. Transphenoidal microsurgery of the normal and pathological pituitary. *Clinical Neurosurgery. Proceedings of the Congress of Neurological Surgeons, 1968. Baltimore: Williams and Wilkins; 1969:185–217.*
109. Nishioka H, Inoshita N, Mete O, et al. The complementary role of transcription factors in the accurate diagnosis of clinically nonfunctioning pituitary adenomas. *Endocr Pathol.* 2015;26(4):349–355.
110. Komor J, Reubi JC, Christ ER. Peptide receptor radionuclide therapy in a patient with disabling non-functioning pituitary adenoma. *Pituitary.* 2014;17(3):227–231.
111. Chandrasekharappa SC, Guru SC, Manickam P, et al. Positional cloning of the gene for multiple endocrine neoplasia-type 1. *Science.* 1997;276:404–407.
112. Zhuang Z, Ezzat S, Vortmeyer AO, et al. Mutations of the MEN1 tumor suppressor gene in pituitary tumors. *Cancer Res.* 1997;57:5446–5451.
113. Asa SL, Somers K, Ezzat S. The MEN-1 gene is rarely down-regulated in pituitary adenomas. *J Clin Endocrinol Metab.* 1998;83:3210–3212.
114. Pellegata NS, Quintanilla-Martinez L, Siggelkow H, et al. Germ-line mutations in p27Kip1 cause a multiple endocrine neoplasia syndrome in rats and humans. *Proc Natl Acad Sci USA.* 2006;103(42):15558–15563.
115. Georgitsi M, Raitila A, Karhu A, et al. Germline CDKN1B/p27Kip1 mutation in multiple endocrine neoplasia. *J Clin Endocrinol Metab.* 2007;92(8):3321–3325.
116. Agarwal SK, Mateo CM, Marx SJ. Rare germline mutations in cyclin-dependent kinase inhibitor genes in multiple endocrine neoplasia type 1 and related states. *J Clin Endocrinol Metab.* 2009;94(5):1826–1834.

117. Kirschner LS, Carney JA, Pack SD, et al. Mutations of the gene encoding the protein kinase A type I-alpha regulatory subunit in patients with the Carney complex. *Nat Genet.* 2000;26(1):89–92.
118. Yin Z, Williams-Simons L, Parlow AF, Asa S, Kirschner LS. Pituitary-specific knockout of the Carney complex gene prkar1a leads to pituitary tumorigenesis. *Mol Endocrinol.* 2008;22(2):380–387.
119. Kaltsas GA, Kola B, Borboli N, et al. Sequence analysis of the PRKAR1A gene in sporadic somatotroph and other pituitary tumours. *Clin Endocrinol (Oxf).* 2002;57 (4):443–448.
120. Sandrini F, Kirschner LS, Bei T, et al. PRKAR1A, one of the Carney complex genes, and its locus (17q22-24) are rarely altered in pituitary tumours outside the Carney complex. *J Med Genet.* 2002;39(12):e78.
121. Beckers A, Daly AF. The clinical, pathological, and genetic features of familial isolated pituitary adenomas. *Eur J Endocrinol.* 2007;157(4):371–382.
122. Vierimaa O, Georgitsi M, Lehtonen R, et al. Pituitary adenoma predisposition caused by germline mutations in the AIP gene. *Science.* 2006;312(5777):1228–1230.
123. DiGiovanni R, Serra S, Ezzat S, Asa SL. AIP mutations are not identified in patients with sporadic pituitary adenomas. *Endocr Pathol.* 2007;18(2):76–78.
124. Denes J, Kasuki L, Trivellin G, et al. Regulation of aryl hydrocarbon receptor interacting protein (AIP) protein expression by MiR-34a in sporadic somatotropinomas. *PLoS One.* 2015;10(2):e0117107.
125. Papathomas TG, Gaal J, Corssmit EP, et al. Non-pheochromocytoma (PCC)/paraganglioma (PGL) tumors in patients with succinate dehydrogenase-related PCC-PGL syndromes: a clinicopathological and molecular analysis. *Eur J Endocrinol.* 2014;170(1):1–12.
126. Gill AJ, Toon CW, Clarkson A, et al. Succinate dehydrogenase deficiency is rare in pituitary adenomas. *Am J Surg Pathol.* 2014;38(4):560–566.
127. Asa SL, Ezzat S. The pathogenesis of pituitary tumours. *Nat Rev Cancer.* 2002;2 (11):836–849.
128. Ezzat S, Asa SL. Mechanisms of disease: the pathogenesis of pituitary tumors. *Nat Clin Pract Endocrinol Metab.* 2006;2(4):220–230.
129. Asa SL, Ezzat S. The pathogenesis of pituitary tumors. *Annu Rev Pathol.* 2009;4:97–126.
130. Newey PJ, Nesbit MA, Rimmer AJ, et al. Whole-exome sequencing studies of nonfunctioning pituitary adenomas. *J Clin Endocrinol Metab.* 2013;98(4):E796–E800.
131. Celli G, LaRochelle WJ, Mackem S, Sharp R, Merlino G. Soluble dominant-negative receptor uncovers essential roles for fibroblast growth factors in multi-organ induction and patterning. *EMBO J.* 1998;17(6):1642–1655.
132. Roh M, Paterson AJ, Asa SL, Chin E, Kudlow JE. Stage-sensitive blockade of pituitary somatomammotrope development by targeted expression of a dominant negative epidermal growth factor receptor in transgenic mice. *Mol Endocrinol.* 2001;15(4):600–613.
133. Scully KM, Rosenfeld MG. Pituitary development: regulatory codes in mammalian organogenesis. *Science.* 2002;295(5563):2231–2235.
134. Itoh N, Ornitz DM. Evolution of the Fgf and Fgfr gene families. *Trends Genet.* 2004;20 (11):563–569.
135. Ezzat S, Walpola IA, Ramyar L, Smyth HS, Asa SL. Membrane-anchored expression of transforming growth factor-a in human pituitary adenoma cells. *J Clin Endocrinol Metab.* 1995;80:534–539.
136. McAndrew J, Paterson AJ, Asa SL, McCarthy KJ, Kudlow JE. Targeting of transforming growth factor-a expression to pituitary lactotrophs in transgenic mice results in selective lactotroph proliferation and adenomas. *Endocrinology.* 1995;136:4479–4488.
137. LeRiche V, Asa SL, Ezzat S. Epidermal growth factor and its receptor (EGF-R) in human pituitary adenomas: EGF-R correlates with tumor aggressiveness. *J Clin Endocrinol Metab.* 1996;81:656–662.

138. Ezzat S, Zheng L, Smyth HS, Asa SL. The c-erbB-2/neu proto-oncogene in human pituitary tumours. *Clin Endocrinol (Oxf).* 1997;46:599–606.
139. Cooper O, Vlotides G, Fukuoka H, Greene MI, Melmed S. Expression and function of ErbB receptors and ligands in the pituitary. *Endocr Relat Cancer.* 2011;18(6): R197–R211.
140. Haddad G, Penabad JL, Bashey HM, et al. Expression of activin/inhibin subunit messenger ribonucleic acids by gonadotroph adenomas. *J Clin Endocrinol Metab.* 1994;79:1399–1403.
141. Penabad JL, Bashey HM, Asa SL, et al. Decreased follistatin gene expression in gonadotroph adenomas. *J Clin Endocrinol Metab.* 1996;81:3397–3403.
142. Abbass SAA, Asa SL, Ezzat S. Altered expression of fibroblast growth factor receptors in human pituitary adenomas. *J Clin Endocrinol Metab.* 1997;82:1160–1166.
143. Zhu X, Asa SL, Ezzat S. Fibroblast growth factor 2 and estrogen control the balance of histone 3 modifications targeting MAGE-A3 in pituitary neoplasia. *Clin Cancer Res.* 2008;14(7):1984–1996.
144. Zhu X, Lee K, Asa SL, Ezzat S. Epigenetic silencing through DNA and histone methylation of fibroblast growth factor receptor 2 in neoplastic pituitary cells. *Am J Pathol.* 2007;170(5):1618–1628.
145. Qian ZR, Sano T, Asa SL, et al. Cytoplasmic expression of fibroblast growth factor receptor-4 in human pituitary adenomas: relation to tumor type, size, proliferation, and invasiveness. *J Clin Endocrinol Metab.* 2004;89(4):1904–1911.
146. Yu S, Asa SL, Weigel RJ, Ezzat S. Pituitary tumor AP-2alpha recognizes a cryptic promoter in intron 4 of fibroblast growth factor receptor 4. *J Biol Chem.* 2003;278 (22):19597–19602.
147. Ezzat S, Zheng L, Asa SL. Pituitary tumor-derived fibroblast growth factor receptor 4 isoform disrupts neural cell-adhesion molecule/N-cadherin signaling to diminish cell adhesiveness: a mechanism underlying pituitary neoplasia. *Mol Endocrinol.* 2004;18 (10):2543–2552.
148. Daniel L, Trouillas J, Renaud W, et al. Polysialylated-neural cell adhesion molecule expression in rat pituitary transplantable tumors (spontaneous mammotropic transplantable tumor in Wistar-Furth rats) is related to growth rate and malignancy. *Cancer Res.* 2000;60(1):80–85.
149. Zhu X, Mao X, Hurren R, Schimmer AD, Ezzat S, Asa SL. Deoxyribonucleic acid methyltransferase 3B promotes epigenetic silencing through histone 3 chromatin modifications in pituitary cells. *J Clin Endocrinol Metab.* 2008;93(9): 3610–3617.
150. Liu W, Asa SL, Ezzat S. Vitamin D and its analog EB1089 induce p27 accumulation and diminish association of p27 with Skp2 independent of PTEN in pituitary corticotroph cells. *Brain Pathol.* 2002;12(4):412–419.
151. Amaral FC, Torres N, Saggioro F, et al. MicroRNAs differentially expressed in ACTH-secreting pituitary tumors. *J Clin Endocrinol Metab.* 2009;94(1):320–323.
152. Bellodi C, Krasnykh O, Haynes N, et al. Loss of function of the tumor suppressor DKC1 perturbs p27 translation control and contributes to pituitary tumorigenesis. *Cancer Res.* 2010;70(14):6026–6035.
153. Ezzat S, Mader R, Yu S, Ning T, Poussier P, Asa SL. Ikaros integrates endocrine and immune system development. *J Clin Invest.* 2005;115(4):1021–1029.
154. Ezzat S, Mader R, Fischer S, Yu S, Ackerley C, Asa SL. An essential role for the hematopoietic transcription factor Ikaros in hypothalamic-pituitary-mediated somatic growth. *Proc Natl Acad Sci USA.* 2006;103(7):2214–2219.
155. Ezzat S, Yu S, Asa SL. Ikaros isoforms in human pituitary tumors: distinct localization, histone acetylation, and activation of the 5' fibroblast growth factor receptor-4 promoter. *Am J Pathol.* 2003;163(3):1177–1184.

156. Ezzat S, Zhu X, Loeper S, Fischer S, Asa SL. Tumor-derived Ikaros 6 acetylates the Bcl-XL promoter to up-regulate a survival signal in pituitary cells. *Mol Endocrinol.* 2006;20(11):2976–2986.
157. Loeper S, Asa SL, Ezzat S. Ikaros modulates cholesterol uptake: a link between tumor suppression and differentiation. *Cancer Res.* 2008;68(10):3715–3723.
158. Dorman K, Shen Z, Yang C, Ezzat S, Asa SL. CtBP1 interacts with Ikaros and modulates pituitary tumor cell survival and response to hypoxia. *Mol Endocrinol.* 2012;26 (3):447–457.
159. Fedele M, Battista S, Kenyon L, et al. Overexpression of the HMGA2 gene in transgenic mice leads to the onset of pituitary adenomas. *Oncogene.* 2002;21(20):3190–3198.
160. Fedele M, Pentimalli F, Baldassarre G, et al. Transgenic mice overexpressing the wild-type form of the HMGA1 gene develop mixed growth hormone/prolactin cell pituitary adenomas and natural killer cell lymphomas. *Oncogene.* 2005;24(21):3427–3435.
161. De M I, Visone R, Wierinckx A, et al. HMGA proteins up-regulate CCNB2 gene in mouse and human pituitary adenomas. *Cancer Res.* 2009;69(5):1844–1850.
162. Finelli P, Pierantoni GM, Giardino D, et al. The High Mobility Group A2 gene is amplified and overexpressed in human prolactinomas. *Cancer Res.* 2002;62 (8):2398–2405.
163. Evans CO, Moreno CS, Zhan X, et al. Molecular pathogenesis of human prolactinomas identified by gene expression profiling, RT-qPCR, and proteomic analyses. *Pituitary.* 2008;11(3):231–245.
164. Qian ZR, Asa SL, Siomi H, et al. Overexpression of HMGA2 relates to reduction of the let-7 and its relationship to clinicopathological features in pituitary adenomas. *Mod Pathol.* 2009;22(3):431–441.
165. Kumar TR, Graham KE, Asa SL, Low MJ. Simian virus 40 T antigen-induced gonadotroph adenomas: a model of human null cell adenomas. *Endocrinology.* 1998;139: 3342–3351.
166. Ruebel KH, Leontovich AA, Tanizaki Y, et al. Effects of TGFbeta1 on gene expression in the HP75 human pituitary tumor cell line identified by gene expression profiling. *Endocrine.* 2008;33(1):62–76.

INDEX

G

U

V

W

Z

Kolibri 360 GmbH
Ferdinand-Jühlke-Straße 7
99095 Erfurt, Deutschland
produktsicherheit@kolibri360.de